FOOD PRODUCTION IN URBAN AREAS

Dedication

I dedicate this book to my late father Daniel Obosu-Mensah (Teacher Mensah of Oyoko, Koforidua) and to my son Konadu and his sisters Nana Adwoa, Natalie and Maria Adowaa

Food Production in Urban Areas

A study of urban agriculture in Accra, Ghana

KWAKU OBOSU-MENSAH

Ashgate

Aldershot • Brookfield USA • Singapore • Sydney

Published by
Ashgate Publishing Limited
Gower House
Croft Road
Aldershot
Hampshire GU11 3HR
England

Ashgate Publishing Company
Old Post Road
Brookfield
Vermont 05036
USA

Ashgate website: http://www.ashgate.com

British Library Cataloguing in Publication Data
Obosu-Mensah, Kwaku, 1958-
Food production in urban areas : a study of urban agriculture in Accra, Ghana
1.Urban agriculture - Ghana - Accra 2.Urban agriculture - Ghana - Accra - History 3.Food supply - Ghana - Accra
I.Title
338.1'09667

Library of Congress Catalog Card Number: 99-72607

ISBN 0 7546 1029 2

Printed in Great Britain by
Antony Rowe Ltd, Chippenham, Wiltshire

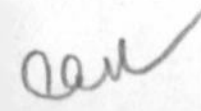

Contents

List of Tables

List of Figures

List of Maps

Acknowledgements

This book started as my Ph.D. thesis at the University of Toronto. Therefore, my sincerest thanks go to my supervisor Professor Bill Michelson and the other members of my thesis committee, namely Professors Michal Bodemann and Richard Stren, who all were very helpful in the successful completion of the thesis. I also thank Professor P.W.K. Yankson of the University of Ghana, Legon, for his invaluable suggestions while in the field. In addition, I sincerely thank my field assistants in the persons of Miss Alberta Indome (Legon), Mr. Danso (Legon), Miss Monica Azinab (IPS), and Obarimaba Akuoko-Marfo of Christ Apostolic Church, Ashiaman. Furthermore, I acknowledge the help of Angela of Ashiaman who assisted me with the data entries. My nephew Stephen Ako-Adjei and niece Nana Asor should also be commended for staying with me in Accra and providing me with assistance of all forms during my field work.

Mention should also be made of the ever helpful Professor Charles Jones, and Jeannette Wright of the Sociology Department of the University of Toronto. Furthermore, this acknowledgement would be meaningless if I did not extend my appreciation to Professor Nancy Howell, who was always helpful to me during my years in Toronto. Thanks very much, Prof. Howell.

I also thank Miss Martina Rowley for proofreading the manuscript, and both Dr. Emmanuel Koku and Dr. Robert Inkoom for typesetting and other types of help that freed some time for me to concentrate on the book. Thanks be to my good friend Dr. Rima Wilkes for her friendship and moral support.

Many people were helpful to me during my era as a Ph.D. student at the University of Toronto in Canada. Those who helped me as friend, and consequently as a source of inspiration in one way or another, know who they are. Since they are too many for me to attempt to list, I only have to say *thank you* to all those unnamed individuals.

Last but not least, I take this opportunity to thank my main sponsor, the Rockefeller Foundation of the U.S.A., for funding my fieldwork in Ghana.

In the sponsorship realm, I should also thank the Centre for International Studies of the University of Toronto for awarding me the Sir Val Duncan Travel Award in 1994, and the School of Graduate Studies also of the University of Toronto for awarding me the Muriel D. Bisell Fund in 1994.

To conclude, I have to state categorically that the usual disclaimer applies to any errors in this book. None of the people mentioned above is liable whatsoever for any statement herein made.

Kwaku Obosu-Mensah
Toronto, Canada
June, 1999

1 Introduction and Overview of the Book

1.1 Introduction

This study deals with an important informal sector activity. Urban agriculture is important, especially since many urban residents in developing countries rely on town/city agriculture in one way or another. One of the most important needs of urban people in developing countries is food. In recent past, food for urban residents came from rural areas, however, due to various reasons the food supply from rural areas is inadequate for many towns/cities in the developing world, especially in sub-Saharan Africa.

In 1981, the World Bank observed that,

> ... for most African countries, and for a majority of the African population, the record is grim, and it is no exaggeration to talk of crisis. Slow overall economic growth, sluggish agricultural performance coupled with rapid rates of population increase, and balance-of-payments and fiscal crises—these are dramatic indicators of economic trouble (World Bank, 1981:2).

Writing on the economic and food crises on the continent, Hansen (1989) confirmed the position of the World Bank by noting that 'it is now easily conceded even by the most optimistic observer that Africa is in the midst of a severe [food] crisis' (Hansen, 1989:184).

More than a decade later, in 1998/1999, one could still make such an observation about stagnant, if not decreasing, economic conditions in sub-

Saharan Africa. As noted elsewhere, in present-day sub-Saharan Africa, 'rural areas do not produce enough food to feed both rural and urban people and importation is constrained by lack of sufficient foreign exchange' (Sawio, 1994:25). The International Monetary Fund (IMF) notes after praising a few African countries for improving their economic conditions that 'in a number of other [sub-Saharan African] countries ... economic conditions remain difficult (IMF, 1996:11).

The seriousness of the poor economic conditions reported for sub-Saharan Africa is compounded by the trend of population growth in that area. Over the years there has been a high population growth coupled with decreased productivity. This trend implies the dire need for programmes to improve upon food production on the continent because 'it is in the field of agriculture that the crisis manifests itself in its most virulent form' (Hansen, 1989:186). It is an uncontroversial fact that in sub-Saharan Africa food sufficiency ratios have dropped since the 1960s. According to an observer, 'by 1980 food self-sufficiency ratios had dropped from 98 per cent in the 1960s to around 86 per cent, which means that each African has on average 12 per cent less grown food in 1980 than 20 years ago' (Hansen, 1989:186).

The situation is not better in the 1990s. According to Svedberg (1991), the per capita availability of food has declined over the 1970s and 1980s and is now below 80 per cent of FAO/WHO recommended per capita intake. In 1994 Pinstrup-Andersen observed that 'many Africans are worse off today than they were a decade ago' (Pinstrup-Andersen, 1994:6). While food sufficiency ratios have dropped, the demand for food has gone up as a result of population growth, especially in urban areas. Food insufficiency in urban sub-Saharan Africa is mostly caused by a combination of factors including, as already noted, inadequate food production in rural areas, faulty government policies, poor distribution and storage facilities, farmer alienation due to the use of crude implements, and a concentration on export or cash-crop production to the detriment of food-crop production.

Food scarcity in sub-Saharan Africa is compounded by increased urbanization and proletarianization, which has removed a significant number of people from the traditional agrarian sector which feeds the population. The World Bank, as well as the United Nations (UN), has noticed a high urban population growth in sub-Saharan Africa. The UN, for example, estimates that between 1995 and 2000 the annual rate of change of the urban population in Africa will be about 4.72 percent per year (United Nations, 1995). In Ghana, the urban growth for 1995-2000 is estimated at 4.62 per cent as compared to 3.0 per cent for the country as a whole (UN, 1995).

Table 1.1 Actual and Projected Urban Population in Some Selected African Countries

Country	Percentage Urban (Year)							
	1950	1980	1990	1995	2000	2005	2010	2015
Angola	7.6	21.0	28.3	32.2	36.2	40.2	44.2	48.1
Botswana	0.3	15.1	27.5	35.0	42.2	48.5	53.5	57.1
Gabon	11.4	35.8	45.7	50.0	53.8	57.3	60.7	63.9
Ghana	14.5	30.7	33.0	35.1	37.9	41.3	45.3	49.2
Kenya	5.6	16.1	23.6	27.7	31.8	35.7	39.7	43.7
Nigeria	10.1	27.1	35.2	39.3	43.3	47.2	51.1	54.8
Senegal	30.5	34.9	38.4	41.1	44.5	48.4	52.2	55.9
Zambia	8.9	39.7	49.9	54.5	58.7	62.4	65.4	68.3

Source: U.N.'s World Economic and Social Survey 1995.

Table 1.1, which is self-explanatory, shows that an increasing number of sub-Saharan Africans are moving into urban areas, thus making it more necessary for programmes that increase food production.

Urbanization influences domestic food production and consumption depending on the region in question. In developed countries, increased urbanization does not necessarily lead to a decrease in the quantity of food produced. This is due to the use of improved agricultural machinery, which makes it possible for a few farmers to produce enough to feed the population. In developing countries, however, most of the cultivators still use crude farm implements which limits the number of acres an average farmer cultivates. Thus, the fewer the number of farmers, the lesser the amount of food produced. Therefore, increased urbanization may lead to food shortages due to the migration of farm labour from the countryside.

The introduction of Structural Adjustment Programmes (SAP) in various parts of sub-Saharan Africa and the attendant labour redundancy, which this has produced, have compounded the precarious economic condition for many people, especially urban residents. Employment and wages play a central role in determining the food security of urban households. Without wages, urban residents may not be able to purchase food. The urban unemployed are therefore in crisis because their sources of cash income and consequently their foremost survival base has been denied them. Unemployment, indeed poverty, leads to insufficient food consumption. As Table 1.2 shows, in developing countries undernourishment is higher in urban than in rural areas.

The above table indicates that urban food insecurity and malnutrition abound in developing countries. In most sub-Saharan African countries, official rural agricultural programmes put in place to alleviate this economic hardship have so far not been successful. Consequently, individual urban dwellers have resorted to their own strategies to feed themselves. One such strategy is urban agriculture.

Urban agriculture is a significant part of the informal sector of the economy of most sub-Saharan countries. To define the concept of informal sector, it is almost always contrasted with the formal sector. The concept attracts different conceptualizations because of 'the lack of a clear theoretical basis for the concept as well as wide spectrum of economic activities that it covers' (Singh, 1994:7). In this work, I use the concept of informal sector in reference to the economic activities with the following characteristics: casualness, easy entry, outside the scope of existing company law or government regulations, small-scale operation, reliance on household labour, and labour intensive. This means the informal sector excludes public sector

Table 1.2 Extent of Undernutrition in Selected Developing Countries

		Undernourished Population (%)			
		< 1.2 BMR**		< 1.4 BMR	
Country	Year	Urban	Rural	Urban	Rural
Brazil*	1974/75	6.6	5.4	13.3	9.2
Egypt	1980/81	5.0	8.2	6.5	11.7
India	1971/72	12.7	12.4	19.5	18.7
Indonesia	1976	15.3	16.6	19.9	21.9
Sri Lanka	1980/81	18.7	14.8	26.2	22.5
Sudan	1978/79	12.0	14.1	17.0	19.8
Thailand	1975/76	24.4	17.1	31.8	24.4
Tunisia	1975	9.2	5.4	16.8	10.4

Source: Hussain, M.A. and Lunven, P. 1989.
* Southern Region.
** Basal metabolic rate (BMR) is 'the energy requirement for the maintenance of metabolic integrity, nerve and muscle tone, circulation and respiration, and so on' (Bender, 1997:144). < 1.2 BMR is a more conservative estimate.

establishments, as well as large-size and commercial establishments in the private sector. The informal sector is a veritable superstructure for the survival of the formal sector of developing economies. As Leys (1975) observes, the existence of the informal sector [to which urban agriculture belongs] is essential for the profitable operation of the formal sector of most national economies. It provides cheap goods and services for poorly paid workers. In addition, it sustains the reserve army of workers for eventual employment by the government and Capitalists. Conversely, the formal sector is necessary for the survival of the informal sector. Therefore, the two sectors 'must not be considered as separate dimensions of national growth, but rather as two closely interrelated facets of a single issue, i.e. the investment and organization of natural resources' (Dettwyler, 1985:426).

Due to the implementation of Structural Adjustment Programmes by many African countries, as mentioned above, formal employment is frozen in the public sector.[1] Consequently, the formal sector is not able to guarantee a minimum level of employment. The informal sector has therefore become 'a welcome outlet both for individual citizens grappling with unemployment and for governments happy to observe that unplanned solutions can arise to fill gaps in planning and public policies' (Stren et al, 1992:29). Also commenting on the employment importance of the informal sector of the [Ghanaian] economy, the Institute of Statistics, Social and Economic Research (ISSER) mentioned that,

> alternative means of employment in the formal sector have not by any means kept pace with the redeployments and increase in the urban labor force, leaving most job-seekers no alternative to the informal economy, and the number of workers in the urban informal sector has mushroomed (ISSER, 1995).

An informal sector activity like urban agriculture provides jobs for many urban farmers, as well as artisans, who provide farmers with implements like

---

[1] 'Public sector' is synonymous with 'government enterprises.' It is used to mean government agencies/public corporations engaged in the provision of specific services and goods. The public sector, as well as large-size industrial and commercial establishments in the private sector, are in the formal sector category of the economy. This means the concept of formal sector is broader than the concept of public sector.

hoes, animal pens, and fences.

In spite of its significance in national economies, most countries in sub-Saharan Africa have no official policy on urban agriculture. In fact, in countries like Ethiopia, Uganda, Cameroon, Zimbabwe and others, farms in towns and cities have often been destroyed and livestock confiscated by the political and municipal authorities in town or urban planning processes. However, there is evidence that such 'harassment of urban cultivators has declined in recent years due to the increasing fluctuation in food supply in urban areas from rural subsistence economies' (Simon, 1992: 82).

Equally important but perhaps far more significant is the lack of foreign exchange to sustain a policy of food imports. Not only has this compelled many sub-Saharan African countries to reduce food imports but it has also encouraged urban residents to produce their own food.

The economic decline in sub-Saharan Africa has also made 'urban agriculture an alternative to cash payments for rising cost of food in the urban areas' (Chimhowu and Gumbo, 1993:111). There is little doubt that urban agriculture is an essential survival strategy for many people. However, for many others, it is increasingly becoming a practical money saving activity. This occurs when, for example, successful urban farming releases money to fund the education of children. The apparent lack of official support for urban agriculture, in spite of its importance in the national economy and in the social lives of urban dwellers is, therefore, regrettable. It should be mentioned that there are some official concerns about urban agriculture. These are addressed in chapter 5.

1.2 Background of the Study

Farming in cities and towns is not a recent phenomenon. The focus of most studies of urban agriculture has been on crop rather than animal farming. Unfortunately, the author is a casualty of this syndrome. Thus, focus is on crop cultivation. Should the author have the opportunity in the future, he will study animal rearing by urban residents.

Most published books on urban agriculture were written by geographers, planners, political scientists, environmentalists, and so on. Consequently, their focus is different from mine. Most of them lay little emphasis on explaining *why* things happen. That is, they make several observations without explaining the reasons why. For example, many researchers have asserted that government officials in sub-Saharan Africa are increasingly condoning urban agriculture but they have not explained why this is so. As a sociologist, it is the author's interest to find out why things happen or why things are what they are. As a result, this work is set apart from others.

Sociologically, the study of urban agriculture is important because, among other factors, it gives social scientists the opportunity to gather information on an important coping mechanism employed by urban dwellers. Similarly, through the study of urban agriculture, social scientists are able to know how people in need depend on their social networks for survival. In addition, by studying the characteristics of urban farmers in a particular town/city, one is able to know the least and best developed areas of that country. The study of the characteristics of urban farmers also makes it possible for social scientists to know who is more likely to engage in urban agriculture. In addition, knowledge gathered through the study of urban agriculture makes it possible for one to project the future supply of vegetables and other crops to urban areas.

In a small way, this study will indicate that when small-scale enterprises controlled by lower class citizens become very lucrative, middle/upper class citizens take control over them. Thus, due to their strong economic positions, the middle/upper class appropriates the innovations of the lower class.

1.3 Significance of the Study

A broader knowledge of urban agriculture will increase the prestige of the practice. A significant minority of the people in Ghanaian towns and cities do not appreciate the importance of urban agriculture, so it is only through exposure this attitude will change. A typical example from personal experience: in 1992, the Canadian Sociology and Anthropology Association organized a conference at the University of Toronto. One of the sessions was on urban agriculture, and the synopses of the panel participants highlighted the importance of urban agriculture. I was sceptical about the practice; in fact I had a negative attitude towards urban agriculture. Therefore, I attended this session with the intention of posing difficult questions to the panelists. My aim was to point out to the panelists and other participants that urban agriculture was bad, and consequently should not be encouraged. However, at the end of the session I was converted. All of a sudden, urban agriculture appealed positively to me. This is how I became interested in urban agriculture as an area of study. My point is, had it not been for the exposure of research findings in urban agriculture, I would probably still maintain a negative view towards this important practice.

My work in Ghana, a West African country, is significant because almost all previous studies on this subject were undertaken in East Africa. This work therefore gives the reader the opportunity to compare and contrast urban agriculture of a typical anglophone West African country with those of East African countries.

Another significance of this work is that it helps the reader's awareness of the dynamics of urban agriculture, and of the characteristics of the people involved in the practice. Through this study it is shown how some new migrants to urban Ghana cope with the necessary adjustments they have to make in order to survive urban life.

Accra was chosen for this study because it is the capital of Ghana. In addition, it is the largest city in the country. Accra is an ancient city, pre-dating colonization. Since it was made the capital of Ghana (then, Gold Coast) in 1877 it has been the focus of the location of public institutions—services, administrative and industrial. Accra is therefore the most attractive city to Ghanaian intellectuals as well as business people, and also to the rural population in general. It is also the largest urban area in Ghana.

Accra is located at the south-eastern part of Ghana, a country that shares boundaries with Togo to the east, Cote d'Ivoire to the west, and Burkina Faso to the north. To the south, Ghana tapers into the Atlantic Ocean or the Gulf of Guinea, as this area of the ocean is known. The exact geographical location of Ghana is between latitudes 5 and 11 degrees north of the Equator, and longitudes 3 degrees west and 1 degree east of the Greenwich Meridian. It covers an area of about 239,000 sq. km. (see Map 1).

Map 1 Ghana: Regional Boundaries

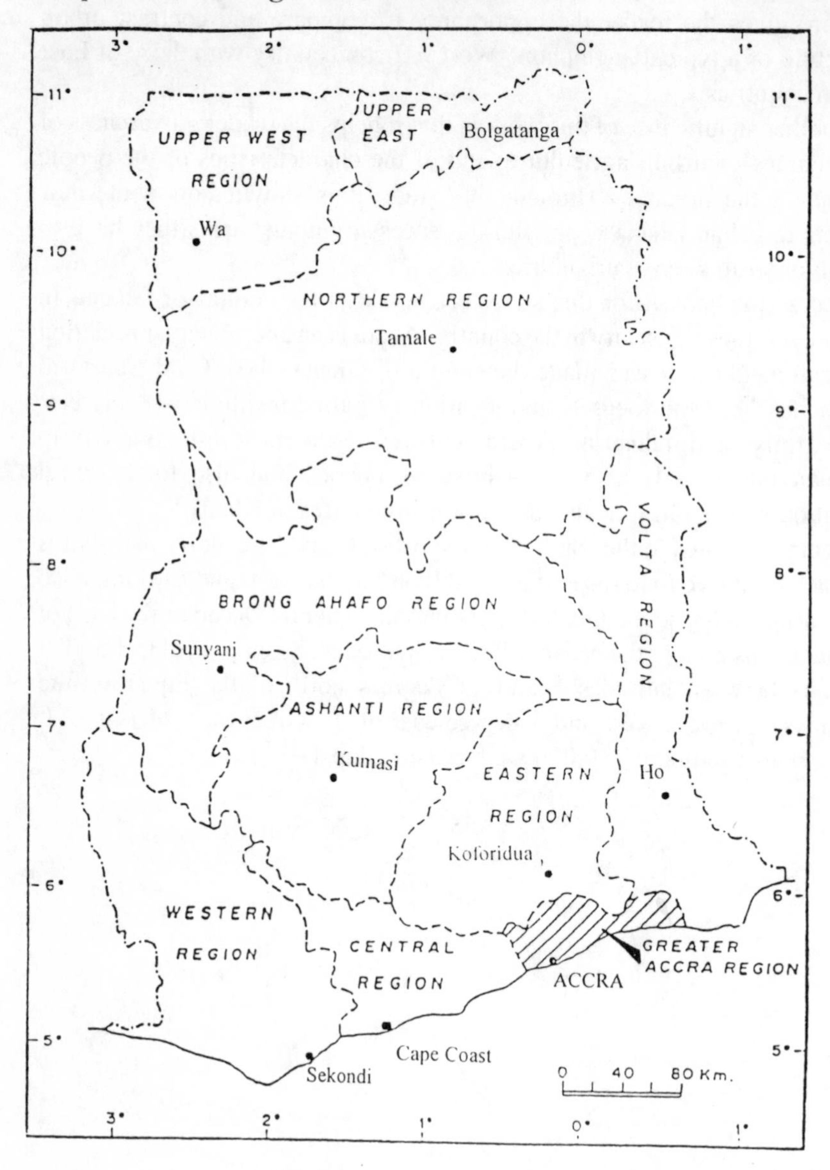

1.4 Definition of the Concept of Urban Agriculture

Urban agriculture (UA) is the main concept of this study. Urban agriculture is simply defined as the practice of farming within the boundaries of towns and cities. Farming in this sense involves crop cultivation, animal rearing, bee keeping, fish farming, etc. In the definition of urban agriculture, it is the location of farms which plays the most important role. A farm should be located in an urban area in order to be called an urban farm. Thus an urban dweller, who farms or maintains farms in a rural area, is not an urban farmer. On the other hand, a person residing in a rural area but farming in a city is an urban farmer.[2]

1.5 Categories or Types of Urban Agriculture

On the basis of location of farms, two main categories of urban agriculture were identified: open-space farming and enclosed farming.

The concept of open-space farming is used to mean farming on land some distance away from premises. The farming is done in the open, on land that does not normally belong to the cultivator.

Enclosed farming, on the other hand, is farming on land close to one's premises. As the name denotes, such lands are within the walls or fences enclosing houses, and normally the plots of land belong to the cultivators.

Open-space and enclosed farmers are quite different, and hold different reasons for farming in urban areas. Similarly, they encounter different problems and consequently adopt different strategies in solving their problems. Unless otherwise stated, the concept of urban farmer or farming used in this work refers to an open-space farmer, reflecting the emphasis of this work.

In chapter 7, the main differences between open-space and enclosed farmers are discussed.

1.6 Overview of the Book

The present chapter gives a short introduction to this book. Here, among other issues, the concept of urban agriculture is defined. The next chapter is devoted

---

[2] The latter rarely happens. There is hardly an urban farmer who resides in a rural area.

to reviewing some current theoretical materials on urban agriculture. It analyses what other social scientists have written on why some urban residents engage themselves in farming. An additional model helps explain one of the most important characteristics of urban cultivators—their rural background.

Chapter 3 is on methodology. It explains where and how the data were collected and analysed. In addition, the limitations of the study as far as data collection is concerned are pointed out.

In Chapter 4, a history of contemporary urban agriculture is discussed, noting when urban agriculture became an issue of importance in Ghana.

Change in official attitude toward urban agriculture is discussed in chapter 5. It shows reasons why some officials had (some still have) negative attitudes toward urban agriculture. Similarly, the factors that influenced many officials to change their attitude toward the practice are discussed.

The characteristics of urban cultivators in Accra are studied in chapter 6. This looks at variables like the socio-economic background of cultivators, their gender and age, the length of time they stayed in Accra before they cultivated, their home origin, and the like.

Chapter 7 analyses the importance of social inequality in determining the type of cultivation an individual is likely to engage in. People of higher socio-economic status are involved in enclosed farming, while people of lower socio-economic status are mostly involved in open-space farming.

Chapter 8 is mainly about the effects of social networks on urban agriculture. Here, the fact that many urban farmers depend on their social network members to acquire land and other such necessities that help them in farming is underlined.

Urban cultivators face some problems that affect their productivity. Such problems include inadequate land tenure system, lack of credit, pilfering, and so on. These problems, and how the farmers try to solve them, are discussed in chapter 9.

The last chapter, chapter 10, is devoted to a short conclusion. In addition, the author predicts the future trend of urban farming in Accra, and on the basis of that and other factors recommends some policies Ghanaian and officials of other sub-Saharan African countries may adopt in order to improve upon the practice.

2 Review of the Literature

2.1 Urban Agriculture: The Recent Debate

This chapter discusses three main issues. The first one is a debate about why urban agriculture in sub-Saharan Africa is on the increase. After establishing this, the reasons why some urban dwellers farm in urban areas and others do not is discussed. In addition, it is argued that contrary to the view of some observers urban agriculture is a permanent feature in sub-Saharan African countries.

Some scholars have pointed out that despite the occasional harassment of urban farmers in some countries, urban agriculture is on the increase (Diallo, 1993; Mougeot, 1993; Maxwell and Zziwa, 1992; Freeman, 1991). This assertion is an indication that urban agriculture has an encouraging future—that it is a permanent issue for those involved and, all things being equal, many more people would be involved in the practice.

There might be various reasons for the rise in urban agriculture but for most researchers the most important cause is the declining purchasing capacity of urban dwellers, both the waged and the non-waged. Table 2.1 shows the purchasing power of the inhabitants of urban Ghana and Kampala (Uganda).

It should be noted that in the table below the part of the Ghanaian data is on urban Ghana rather than on Accra or any other specific town/city. It should be mentioned that in one study, Alderman and Shively (1996) found that the minimum wage purchased less food in Accra than in any other town or city in Ghana. This means the purchasing power of the inhabitants of Accra is lower than the average reported in Table 2.1. In any case, figures in the table indicate that the purchasing power of people living in Kampala and those living in urban Ghana has decreased since the mid 1970s. In 1992, a similar conclusion

Table 2.1 Purchasing Power: Wage/Price Ratio in Urban Ghana and Kampala

Year	1970	1972	1974	1976	1978	1980	1982	1984	1986	1988	1990
Urban Ghana	100	103	149	74	39	22	19	18	34	35	49
Kampala	100	100	--	32	--	5.4	7.2	8.1	5.4	6.0	10

Source: Jamal and Weeks 1993; ISSER 1995.

was made that in Uganda 'wage income fell precipitously in relation to the cost of living between the end of the 1970s and the present...' (Bigsten and Kayizzi-Mugerwa, 1992:1432).

A drop in real wages has been reported in other sub-Saharan African countries. For example, in Nigeria 'within the past decade the real wage decline assumed crisis proportions... While other prices soared in response to the rationalisations of SAP (Structural Adjustment Programme), the price of labour was artificially held in check' (Ikpeze, 1994:72).

The drop in purchasing power can be partly attributed to increased multi-national corporations, increased demand for higher pay in core countries, higher operational costs, and corruption within government (political) and state functionaries of sub-Saharan African countries. The increased demand for higher pay in core countries has tended to increase the prices of commodities imported by developing countries, thus lowering the purchasing power of workers in such countries. It has also increased the amount of transfer from developing countries to developed countries in the form of debt servicing, and payment for imports. As noted elsewhere,

> Outflows of dividends, royalties, management fees, etc exceed inflows of new investment and are financed from commodity exports, while the investment in manufacturing is generally for the domestic market and absorbs rather than earns foreign exchange (Rakodi, 1997:45).

Among other things, this compels developing countries to offer lower wages to their workers, thus decreasing the workers' purchasing power. Furthermore, money which could have been used in developing countries is transferred out of such countries by multi-national corporations as profit. In this case, 'money capital [is] invested in production in the periphery, and the profits derived from the sale of the materials produced (either raw materials or goods for import substitution) [are] repatriated to the home base' (Forbes, 1986:80). So, 'unequal exchange allow[s] a transferral of surplus value internationally via the sphere of circulation' (Gibson, 1980:171). In addition, by underpaying workers and increasing costs of health care, education and the like, corrupt officials in developing countries deny workers their fair share of money generated, thus further decreasing workers' purchasing power.

Another reason for the increase in urban agriculture is that 'urban agriculture is, relatively speaking, potentially lucrative' (Dettwyler, 1985:248). It is lucrative, especially for those who cultivate for sale, partly because of decreasing purchasing power, which has made it a necessary supplement for wage income for many urban residents. Generally, it seems 'the risks from harassment and destruction of crops by authorities, loss through theft and predation, and other drawbacks to cultivation of urban open space are outweighed by the perceived advantages and gains from cultivation, since the practice is spreading in the cities' (Freeman, 1991:20).

Similar to Freeman's position is the assertion that 'as economies [in sub-Saharan Africa] have failed, urban agriculture has become an alternative to cash payments for food in the urban areas' (Chimhowu and Gumbo, 1993:111). Instead of buying all their foodstuffs, some urban residents produce some of their own food, and save money for essential non-food items like school fees, health care and shelter. On the question of why some urban residents engage in urban agriculture and others do not, a number of researchers have attempted to provide answers. These are epitomized by Freeman's (1991) two models. The first model, the labour-surplus model, is based on concepts of competition in a labour-surplus economy. This model stipulates that due to surplus labour there is a fierce competition for limited jobs, and consequently the less qualified are rendered jobless. Subsequently, the poor and the unemployed engage in urban farming for their survival. The model helps to explain why in some countries urban agriculture is dominated by unschooled female labour, people who are forced to concede the better paying formal sector jobs to the more highly educated males, who are plentiful in 'the reserve army of the urban unemployed' (Todaro, 1981). An implication is that most of the jobless, who form the majority of the reserve army of labour, view their involvement in urban agriculture as a temporary situation. Their involvement in urban agriculture is a temporary issue because they hope to find salaried jobs in the formal sector, and subsequently abandon farming. As far as this model is concerned, the economic survival of the farmers is dependent upon urban agriculture because during the period of farming, most of them are not formally employed, 'they don't have a choice, they don't have any source of income' (Maxwell, 1993:10). In this context, urban residents' involvement in urban farming is therefore a survival strategy—they may not survive urban life without farming. The labour-surplus model indicates that there are no alternative, more remunerative activities available for many urban residents so they are compelled to engage in farming. In Freeman's terms, urban farming is part of the self-maintenance of the unemployed, removing from employers or, indeed, from the government the burden of maintaining this potential labour force in the city. The crux of the model is supported by findings in some sub-Saharan African countries. Lamba (1993) notes that many of the urban farmers in Nairobi (Kenya) said they would starve if they were not farming. According to Maxwell (1994), for a group of urban farmers in Uganda, farming constitutes a survival strategy. A study in Ethiopia by Egziabher (1994) demonstrates that many urban people engage in urban farming as a matter of survival. Since women are particularly disadvantaged in the competition for the scarce formal employment opportunities in urban areas, it is presumed that in the context of this model most urban farmers are women. Actually, the studies conducted in Kenya by Freeman, and the one conducted by the Mazingira Institute (also in Kenya), showed that about two-thirds of the respondents were female. In Uganda, Zambia and Ethiopia researchers have shown that most of the cultivators are female.

The second model used by Freeman to explain some people's involvement in urban farming is the dependency model of Third World economies. It presumes that the urban economy is characterized by a series of unequal economic and social exchanges between the capitalists and the proletarians. The former appropriate and utilize the surplus value generated by the latter, so the latter become marginalized and poor to the extent that their income cannot sustain them. Consequently, they engage themselves in urban farming to supplement their income. A corollary of this model is that a large number of urban farmers are employed in the formal sector. However, they receive very low wages on which they cannot survive. In Nairobi, Freeman (1991) observed that over half [of the farmers] continue to hold down a full-time or part-time non-farm job in the city, while carrying on with the cultivation of their urban vegetable plot. As Mougeot (1993) puts it, they have other part or full-time jobs. Yeung (1993) also notes that urban farmers supplement their earnings with income from other activities, such as factory work. In a study of urban farmers in Kampala, it was observed that most urban farmers 'have income from jobs, and they consider it a priority to preserve that income for non-food expenditures' (Maxwell, 1993:10).

Two central themes emerge from this literature. First, from the labour-surplus model it can be deduced that urban farmers are mostly illiterate, female, and formally unemployed. Second, the dependency model, on the other hand, tells us that urban farmers may be formally employed but poorly paid. Thus, some urban farmers are 'semi-proletarianized peasantry': wage earners 'only marginally or partially proletarianized as, over their life cycle, they derived the bulk of the means of subsistence for their families from outside the wage economy' (Arrighi and Saul, 1973:69).

The tenets of the labour-surplus model and the dependency model are not contradictory but they are rather a supplementary contribution to explaining why some urban residents farm and others do not farm in urban areas. The information they convey is that a large number of urban farmers are weak in the competition for available jobs, so they are mostly unemployed, and when employed they are poorly paid. This assertion suggests that the prestige of urban farming is low—it is done by the unemployed and the lowly paid.

Some fundamental questions the above models fail to address are whether proletarians will abandon the practice if their income level becomes high enough to make food and other necessities affordable to them; whether urban farming will go down if education, health, and other services are rendered free of charge by the government: that is, if the average individual does not need extra money for non-food expenses?

It seems reasonable to conclude that if urban agriculture is basically a survival strategy, as assumed by the labour-surplus model, then middle and high-income groups (people who may not need this strategy to survive) would not be involved in the practice.

However, available literature shows that many middle and high-income groups are involved in the practice of urban agriculture. Wekwete (1993) notes that evidence has been provided to show that urban agriculture does not benefit the poorest of the poor, because the poor have no access to credits and land, as a result of both formal and informal gate-keeping processes in the city. In Tanzania, Sawio (1994) found that high-income earners in government positions are involved in urban agriculture. The present work shows that in Accra many middle and high-income groups are involved in urban agriculture (see chapter 6). Other researchers have come to the same conclusion (Mvena et al, 1991; Bongole, 1988; Mtwewe, 1987; Mvena, 1986).

Furthermore, on the arguments of the labour-surplus model one should expect urban agriculture to be practised mostly by recent migrants, because they are less likely to be in well-paying formal employment. This is because (apart from officials transferred to occupy specific positions) as new migrants, they are more likely to be in lower level positions. However, data show that urban farmers are mostly not recent migrants (see chapter 6).

There is no doubt that the degree of involvement in urban agriculture is influenced by diminished purchasing power of urban dwellers. Without doubting the popular assertion that urban agriculture is a survival strategy, I think the need to survive is not the most important factor influencing urban residents when they decide to engage in farming. If even economically viable urban dwellers engage themselves in urban agriculture, and if urban dwellers will still be involved in cultivation even if social services are rendered free of charge, then some other factors make urban agriculture attractive to them.[1]

My point is that there are many sectors or practices in the informal economy, so urban farmers could have chosen other practises like carpentry, masonry and dress-making over urban agriculture. One may argue that the most important need of many urban residents is food, so it is natural for them to engage in farming instead of other informal practices in order to earn money to buy food. That is, why should they buy food when they can produce it themselves? However, it is known that one need be a farmer in order to have access to food. The entitlement theory stipulates that with money or capital people are able to command food. According to this theory, 'people establish command over food in many different ways' (Sen, 1990:34) meaning 'access to food is not only a function of food supply, but is influenced by a variety of factors that affect the capacity of particular households and social groups to establish entitlement over food' (Sobhan, 1990:79). Included in these factors is 'the capacity of households to exchange what they have to sell (e.g. labour power or farm products) for goods and services in the market' (Sobhan, 1990:79). Thus, urban residents don't have to be farmers in order to get access to food. It is obvious that many people who are not farmers consume better or more nutritious food than some farmers.

The labour-surplus model and the dependency model discussed above explain why some people get involved in informal sector activities. However, they do not explain why they choose urban agriculture instead of other informal sector activities. This important issue of choice of informal sector practice brings me to the model I have developed to explain why some urban dwellers get involved in agriculture and others do not.

I have termed this model the cultural lag model (the CL model). By this I seek to argue that for most people, urban agriculture is a cultural practice. Thus, most of the urban residents migrated from rural areas; most of them were farmers or associated with farming in their former places and they find it difficult to abandon farming even in the cities. The point is that, considering the strong relationship between rural and urban areas, between people in the formal sector and those in the informal sector in sub-Saharan Africa, any thesis that ignores cultural influences is bound to have problems. Wherever African rural people dwell, they want to cultivate. This is evidenced in the commentary made some decades ago by Orde-Browne which still holds true today:

---

[1] During my fieldwork, urban cultivators in Accra said they would not stop farming even if social services were provided free of charge.

> The creation of large industrial centres with workers completely divorced from food production would be an entire innovation of very doubtful desirability; it appears most unlikely to occur. The African man, and still more the woman, is firmly attached to the soil, and the whole fabric of social organization is based upon the right to cultivate; it thus seems probable that the native will always aim at having his own home among his own crops, whether in a distant village or as a 'squatter' on an estate (Orde-Browne, 1926:72).

In Cameroun, Ngwa (1987) notes that the farmers in Buea are people who had a previous farming background and who therefore undertook farming in and around the town. In Sudan,

> The majority of those who migrated to Khartoum from rural areas were previously working in agriculture either as self-employed or paid agricultural labour or unpaid family labour (Ibnoaf, 1987:12).

Skolka (1987) for example, notes that the informal sector in developing countries is subsistence activities transferred from rural areas to large cities. In Mali, a similar observation for urban agriculture has been made. According to Dettwyler (1985) 'Senoufo migrants are first and foremost farmers' (Dettwyler, 1985:248). Still in Mali, the assertion that most urban farmers are former rural dwellers is also supported by Diallo (1993) who, in answering his own question, 'Urban farming undoubtedly creates jobs, but at what costs?', noted that the countryside in Mali is emptying out. It can be inferred from this statement that most of urban farmers in Mali migrated from the countryside where they were farmers.

In Nairobi, Freeman (1991) observes that 'the majority of cultivators are migrants from other, mostly agricultural parts of Kenya and neighbouring countries' (Freeman, 1991:57). Others were born in the city (Nairobi) but they 'were farming their present plots before this peri-urban zone was incorporated into the city, and have simply continued this activity as the built-up area expanded around them' (Freeman, 1991:67). The results of the research conducted by the Mazingira Institute support the assertion that most urban farmers are former rural dwellers or have previous experience in farming. This study mentioned that 'urban farmers are either peasants surrounded by urban expansion, or migrants who farm their backgrounds or vacant land' (Lee-Smith and Lamba, 1991). A study of the workers in Kaduna, Nigeria, also showed that 'a very large section of the workers had previously worked as farmers themselves, and as many as 46 per cent of those interviewed ... stated this as their only previous work experience' (Andrae, 1992:206). In Ethiopia, Egziabher (1994) notes that,

> Urban agriculture is a traditional practice... The urban-based population is used to keeping cattle, sheep, and chickens, or growing rainfed crops such as maize and vegetables, on the plots adjacent to their houses (Egziagher, 1994:87).

The cultural lag model is further supported by Mbiba (1995) who notes that, 'in most African cities, urban agriculture is a transfer of rural subsistence agriculture to urban areas' (Mbiba, 1995:2).

As far as the cultural lag model is concerned, my assertion is that urban agriculture is mostly practiced by people who are skilful in agriculture because of their rural roots and previous exposure to agriculture.

The central concept of this model is *skill.* People have different skills, so when they are choosing jobs, they choose those that are suitable for their skills. Thus, farming is not the last resort when they choose informal sector jobs. This explains why in Ghanaian towns and cities, most of the keepers of animals like goats and sheep are from the northern part of the country, where animal rearing is an important tradition. In Kampala (Uganda), Oloya (1988) found that most of those who keep sheep are Bantus from traditional animal rearing areas of eastern and northern Uganda. Conversely, urban residents without rural backgrounds are less likely to engage in urban agriculture. They may prefer other informal practices like carpentry, driving, baking and dress-making to farming. That a person's decision to get involved in urban agriculture is in the final analysis influenced by his/her cultural background may also explain why most urban farmers in East Africa are women—food cultivation for consumption is traditionally women's duty. During colonization, this was reinforced by the colonists' forcing of the men to produce cash crops.

Educational standards in the rural areas of sub-Saharan African countries are very low compared to those in urban areas. Thus the illiteracy rates in rural areas are higher than in urban areas. See the illiteracy rates in the table below.

Table 2.2 Percentage of Illiterates (Age 15+): Selected sub-Saharan African Countries

Country	Urban		Rural	
	Male	Female	Male	Female
Ethiopia (1984)	21.2	42.9	74.4	91.2
Ghana (1997)*	27.8	43.1	39.1	50.8
Guinea (1983)	10.0	27.6	28.2	60.7
Mali (1988)	45.2	66.4	81.7	95.0
Namibia (1991)	9.9	10.6	30.5	34.5
Sudan (1993)	22.3	44.0	44.1	71.8
Togo (1981)	24.8	60.1	65.0	89.0
Uganda (1991)	11.8	22.5	34.7	59.7

* The Ghanaian data excludes Accra, which has the highest rate of literacy in the country.

Sources: 1. Ethiopia, Guinea, Mali, Togo and Uganda: Unesco's 1995 Statistical Yearbook. 2. Namibia and Sudan: Unesco's 1996 Statistical Yearbook. 3. Ghana: computed from Women in Agriculture in Ghana by Beatrice Akua Duncan.

In all the six countries reported in the above table, illiteracy rates are higher (more so for females) in rural areas than in urban areas. Due to their lower educational standard, rural people don't acquire enough skills to be easily absorbed in the formal sector. The implication is that the skills most rural migrants bring with them to urban areas might not be adequate to secure them any high-salaried formal jobs, so they are more likely to engage in an informal job like urban agriculture. A related issue is whether urban dwellers' involvement in urban agriculture is a temporary or permanent practice.

2.2 Permanency of Urban Agriculture

Any activity that has the capacity to grow is a potentially permanent activity. Urban agriculture is a growing phenomenon. It is 'increasingly widely practiced, and its efficiency is continually improving...' (UNDP, 1996:8). Studies in 'Kinshasa, Kampala and Maputo speak of massive shifts of urban land from open space, institutional and transportation use to agricultural production' (UNDP, 1996:25). The preceding statements can be interpreted to mean that for the people involved, urban agriculture is a permanent activity. This assertion is supported by Braun, McComb, Fred-Mensah and Pandya-Lorch (1993) by noting that urban agriculture has become a permanent feature throughout the cities of the developing world, despite being actively discouraged for reasons ranging from aesthetics to ideology. Unfortunately, though cultivation in cities is as old as the formation of cities, and the cultivation of crops and keeping of livestock are widely practiced in sub-Saharan African towns and cities, some people consider urban agriculture as a temporary and irrelevant practice. The critics, who consider urban agriculture ephemeral, have doubted 'whether the informal sector [including urban agriculture] or petty commodity enterprises have a capacity for growth, and thus make an economic contribution or whether they are basically parasitic and transitory' (Moser, 1984). According to Lee-Smith and Memon (1994), urban agriculture is seldom considered worth examining within the context of this debate because it is assumed to be a subsistence activity. The school of thought that doubts the potentials of the informal sector does so because, among other factors, it doubts the potentials of this sector in absorbing the pool of the unemployed.

Despite the above-mentioned reservations, no research finding seeks to assert that urban agriculture, like other informal sector activities, is a temporary issue. As emphasised by the International Labour Organization (ILO),

> The recession of the 1980s and the adjustment policies followed in many developing countries, have resulted in its [i.e. informal sector] expansion as modern sector enterprises, and especially the public sector which is a large employer of labour in most developing countries, were obliged to shed labour or reduce wages drastically. The workers affected, having no employment, have had no alternative than to resort to the informal sector (ILO, 1991:10).

It is specifically noted that urban agriculture 'will grow even larger with projected population growth' (Lee-Smith and Memon, 1994:73). The contraction of the economy in developing countries may continue for a long time. In many countries the pace of urban growth is outstripping the capacity of the formal sector economy to provide jobs. Consequently, the informal sector including urban agriculture would have to absorb many more people in the years ahead. The question is whether it would be able to provide adequate job opportunities for so many people. This depends on the environment in which the sector operates, particularly on the encouragement, or otherwise, of the various African governments.

Labour participation in the informal sector is 'almost certainly increasing on both a full- and part-time basis and at an unprecedented rate' (Gibbon, 1993:14). According to the sponge hypothesis of the informal sector maintained by the International Labour Organization (ILO), the capacity of the informal sector for labour absorption is limitless if the state is able to create the necessary enabling environment for it.[2] The ILO also believes that given the fallout from economic reforms in developing countries, there is no cause to assume that the informal sector is a temporary phenomenon that will disappear in the foreseeable future (ILO, 1991). In this perspective, the informal sector is viewed as a labour sponge with an almost infinite capacity to absorb those who are at any time excluded from formal sector activities. To support its argument that the informal sector is an important employment avenue, the ILO came out with some important figures. It noted that,

> A quarter of the African labour force which is growing at the rate of 6 per cent is urbanised, and that two-thirds of all the wage employment which has a limited rate of increase (2 per cent) per annum is also urban based (ILO, 1991b:42).

The ILO cautioned that given the poor employment situation, a staggering 93 per cent of all additional jobs in urban Africa will need to be generated in the informal sector. This is because in sub-Saharan Africa, the formal sector is an inadequate employer. See Table 2.3.

In every country reported in Table 2.3, the formal sector is not able to provide adequate employment. The average percentage of people in employment in those countries is 8.3. This means only 8.3 per cent of total employment in the nine countries is wage employment. The remaining 91.7 per cent seek employment in the informal sector. The serious inability of the formal sector to provide employment for the population becomes clearer when one considers the fact that in five of the nine countries reported, i.e. Ethiopia, Sierra Leone, Tanzania, Togo and Uganda, the urban formal sector employment ratio is below the average. Actually, evidence from other sub-Saharan African countries portrays the same situation (Vandemoortele, 1991).

---

[2] Here, the capacity of the informal sector to absorb labour is compared to the capacity of the sponge in absorbing water.

Table 2.3 Wage Employment in Selected sub-Saharan African Countries—1990 (thousands)

Country	Total Labour Force	Wage Employment	% in Wage Employment
Ethiopia	15287	362.1	2.4
Ghana	4073	461.0	11.3
Kenya	5996	1005.8	16.8
Madagascar	3552	337.7	9.5
Nigeria	27981	2722.1	9.7
Sierra Leone	1184	69.9	5.9
Tanzania	8174	636.4	7.8
Togo	948	73.4	7.7
Uganda	5239	362.8	6.9
Total	72434	6031.2	8.3

Source: Vandemoortele, 1991:93.

The data in table 2.3 were collected in 1990, before many sub-Saharan African countries embraced the Structural Adjustment Programmes that necessitated retrenchment, and in many countries, the freezing of public sector employment. Therefore, it may be assumed that wage employment has further decreased since 1990. As noted by a commentator, 'on average, the urban unemployment rate has doubled over the past 15 years, rising from 10 per cent in the mid-1970s to about 20 per cent today [1991]' (Vandemoortele, 1991:94). This is an indication that the informal sector is indispensable, as far as urban employment is concerned.

In the face of an overwhelming unemployment, rate what kind of adjustments do the unemployed have to make in order to deal with the situation? Many have lowered their reservation wage and are willing to take up jobs that attract qualifications lower than those they possess. On the basis of the fact that in sub-Saharan African countries most formal employment is provided in the public or modern sector, it seems the most important adjustment made by the unemployed (including many university graduates and other highly educated people) is to seek employment in the informal sector. Contrary to the general situation in industrialized countries, where lack of education and training constitutes the main characteristics of unemployed youths, in many instances educated youths in sub-Saharan Africa are more prone to unemployment than uneducated youths. First and foremost, the formal sector, where the educated mostly seek employment, has stagnated in many countries in the region. The unemployment situation among educated youths gives the impression that unemployment in Africa is positively correlated with the level of education. For example in Nigeria, Fashoyin (1994) has observed that the familiar structure of unemployment, whereby unemployment hardly existed among university graduates, has changed. He notes that during the past decade there has been an increasing number of university graduates experiencing unemployment. Further, 'another group that has become the victim of the unemployment phenomenon in the 1980s includes skilled and experienced workers, who lost their jobs as a result of contractions in the economy' (Fashoyin, 1994:v). The same observation has been made in other sub-Saharan African countries. Writing about the growing unemployment rate within university graduates in Kenyan cities, Vandemoortele (1991) noted that,

> The results of the survey clearly indicate that their [university graduates] employment prospects are worsening over time. The growing length of job search, the increased reliance on government employment, the rise in temporary employment and the reduced utilisation of acquired skills all suggest that their labour market situation has deteriorated since the late 1970s (Vandemoortele, 1991:98).

This leads me to suspect also a high or growing unemployment rate among university graduates in other sub-Saharan African countries including Ghana. Unfortunately, there are no available data to confirm this assertion. Apart from unemployment as a factor that encourages urban residents to engage in urban agriculture, the current privatization and mechanization drive should also be considered an important determinant. In sub-Saharan Africa, privatization goes hand in hand with foreign investments. And foreign investment is increasingly introducing improved technology and sophisticated division of labour in the production sector. Since technological division of labour at the work place makes the work process simpler, it reduces the need to hire skilled labour. Consequently, the capitalists hire mostly unskilled labour, which attracts low salaries. Workers' low salaries compel them to supplement their income through urban agriculture and other informal sector activities.

It should be clear from the discussion so far that though there is overwhelming evidence of unemployment in sub-Saharan African countries, 'it does not mean that those who fail to secure a job in the modern sector all join the ranks of the unemployed' (Vandemoortele, 1991:99). As noted earlier, virtually all of them find employment in the informal sector of the economy. According to Oni (1994), the characteristic features of the informal sector, e.g. easy entry, small initial capital or investment, limited skill, and the labour intensive mode of production would further enhance the future capacity of the sector to absorb more labour.

As far as employment and consequently income are concerned, urban agriculture is an important aspect of the informal sector. It has been noted that 'in several economies, particularly developing ones, it [urban agriculture] is one of the largest urban productive industries' (UNDP, 1996:3). In Ibadan (Nigeria) for example, urban agriculture is the most important informal sector activity which university graduates are entering. This may be true for other cities in sub-Saharan Africa. See Table 2.4 for data from Ibadan.

As the table shows, and as pointed out by the cultural lag model, urban agriculture is not the last resort, meaning many graduates are getting involved in urban agriculture instead of other informal sector activities. It is not necessarily the most lucrative activity in the informal sector but they get involved in it because of their skills and/or their previous knowledge of farming. Actually, 'many Nigerians view farming as a low priority, not the sort of work to be taken by anyone with an education' (Binns, 1994:135). The average Ghanaian has the same mentality. In Ghanaian as well as Nigerian first and second cycle schools, pupils who break school rules receive punishment in the form of clearing weeds. Consequently, by the time students become adults, they have a negative feeling about farming because they see it as a form of punishment. The increasing entry of highly educated and skilled personnel into urban agriculture has brought about an important and interesting development. It has led to an increase in the technological know-how of people involved in the activity. This is because the highly educated and skilled personnel bring with them their know-how. This in turn has resulted in an increase in technological sophistication of urban agriculture.

Table 2.4 Distribution of Graduates in Informal Sector Activities in Ibadan—1989

Activity	Number Involved	%
Agriculture	50	27.8
Sales	28	15.6
Self Employment	4	2.2
Teaching, Coaching	11	6.1
Unskilled Labour	9	5.0
Others	11	6.1
No Response	67	37.2
Total	180	100.0

Source: Adejumobi 1992:12.

Another benefit of the entry of highly educated and skilled people into urban agriculture is the increase in financial investment in the sector. This is more so because many of the new entrants were formerly employed in the formal sector, and lost their jobs due to the current restructuring or Structural Adjustment Programmes taking place in sub-Saharan African countries. Such people entered into urban agriculture with their severance benefits.

The entry of highly educated and skilled people into urban agriculture has also raised the prestige of the practice (see chapter 5).

With an increasing number of people entering urban agriculture, and those already involved not exiting from it, the projection is that the practice is a permanent one for those involved. It is therefore necessary for the various sub-Saharan African authorities to formulate policies that would create a favourable environment for the improvement of urban agriculture.

2.3 Summary

This chapter gave a general idea, from the literature, of the type of people more likely to be involved in urban agriculture. Thus, it tells the reader the factors that compel some people to get involved in urban agriculture. From this chapter, one is able to understand why some urban dwellers choose urban agriculture rather than other informal sector activities. The labour-surplus model, the dependency model, and the cultural lag model are complementary in explaining why some urban residents get involved in agriculture.

This chapter may guide policy makers, when including urban agriculture in city planning. Thus, when planners know of the background of migrants into urban areas of Ghana, they would be able to plan towns and cities to accommodate their interests in urban agriculture. In this chapter, I argued that urban agriculture is an important and permanent practice.

The literature arguments in this chapter are very pertinent for the main body of this work because it helps to explain some of the findings or conclusions in succeeding chapters. Consequently, occasionally they would be referred to. I should also mention that the labour-surplus and the dependency models influenced my interest in urban agriculture. When I detected that they lay less emphasis on the rural background of urban farmers, I decided to show that the cultural background of urban farmers is necessary for understanding the reasons why some urban residents prefer urban agriculture to other urban informal sector activities.

Data collection procedures are very important in any work that involves a field study. As a result, I discuss the issue of data collection in the next chapter.

3 Research Methodology

3.1 Introduction

My particular interest in urban agriculture in Ghana was based on two main reasons. Firstly, Ghana is my birthplace. Secondly, I am not aware of any published work on urban agriculture in Ghana. As a result, I decided to study this practice in Accra to make it possible for those interested to compare and contrast urban agriculture in Accra with that practiced in other sub-Saharan African towns/cities, where research on urban agriculture has been conducted. Sociologically, I became interested in this subject owing to a number of factors. In the first place, the part played by 'social inequality', when a cultivator decides the location of his/her farms. Thus, I am interested in how an urbanite's home region influenced his/her choice of farm location. Also, to study how new migrants adjust to urban life as far as urban cultivation is concerned. Furthermore, finding out the factors which compelled government officials to change their attitude towards urban agriculture made it necessary for me to study this practise.

I used mostly an inductive approach for this work, in that most of the assertions or conclusions made were empirical—induced from the data collected or from rational thinking, and from already existing findings of other researchers. For example, after studying the background of urban cultivators in Accra, I was able to conclude that urban cultivation is mostly done by people who were born and/or once lived in rural Ghana. Similarly, I was able to induce from the data collected the reason why government officials have changed their attitude toward urban agriculture.

3.2 Research Design

It is clear from Map 1 that the Greater Accra Metropolitan Area (GAMA) is the smallest of the ten regions in Ghana. However, it is 'the major industrial centre of Ghana' (Benneh et al., 1993:viii), containing 10 per cent of the country's population (Benneh et al., 1993).

As shown in Map 2 below, GAMA comprises of three districts, namely Accra District (Accra Metropolitan Area), Tema District, and Ga District. In size, Accra District, where my study was conducted, is the smallest of the three districts. However, it is the most densely populated. According to Benneh et al. (1993), about 75 per cent of the population of GAMA is located in Accra District. As the name denotes, Accra—the capital of Ghana—is in Accra District.[1] Most of the people in Accra District live in and around Accra (the core area of this study).

In Accra District, there is a dual structure as far as the built-up areas are concerned. The middle/upper class live in areas like Osu, Cantonments, Teshie/Nungua Estates, Airport Residential Area, and Legon.[2] Of late, affluent residential areas are springing up in and around Batsonaa, Madina, Adenta, Taifa, McCarthy Hill, and many other peripheral areas.

Areas like Accra, Kaneshie, Abossey Okai and Nima—where the bulk of the people lives—are overcrowded. Thus, farm land is less accessible at such areas. Most of the industries in Accra District are located in Accra, Kaneshie, Abossey Okai, Asylum Down and other areas dominated by the poor. So, as expected, the poor live at the least preferred areas of Accra. Similarly, Kaneshie, Abossey Okai, Asylum Down and particular Accra are the most commercial, administrative, and business areas of the District.

For the purposes of this work, Accra comprises the area lying between the four military/police barriers, and the shores of the Gulf of Guinea in that area. One of the barriers (the Nungua barrier) is mounted at the boundary between Nungua and Sakumono at the south-eastern side of Accra. Another one, the Adenta barrier is at the north-eastern part of the city, and yet another one is at Taifa (north-western part). The last barrier known as Weija barrier is mounted at the south-western part of the city.

---

[1] In Ghana, the average Ghanaian knows the whole of Accra District simply as Accra.

[2] Many low class people live in Osu and Cantonments but at a distance from the areas where the middle/upper class live.

The military/police barriers appropriately demarcate Accra because they are erected at the outskirts of the city. The history of these barriers is not documented but I assume they were erected by the first military government of the country (1966-1969) to prevent military troops from outside Accra moving in to seize power of government. The military unit that executed the coup was based in far away Kumasi, meaning they moved in from outside. During that reign, an attempted coup to unseat the government was perpetrated by a military unit from Ho, a town which is also located several kilometers from Accra. Thus to prevent outside military units from moving into Accra to unseat the government they erected military/police barriers at the outskirts of Accra to control all traffic moving into, and later out of, the city. Apart from these military/police barriers, there are no clear boundaries between Accra and the surrounding areas.

Map 2 shows the areas of my field study. In this study, Cantonments, Dzorwulu, Nima, Kaneshie, Abossey Okai, and Osu represent the core area of Accra. The periphery comprises of Teshie, Nungua, Batsonaa, Madina, Legon, Achimota, Taifa, and Weija. After deciding on the areas to study, my research assistants and I went round for feasibility studies. This gave us the dynamics of urban cultivation in the areas, and consequently, an idea of the number of people to be interviewed in each area. During these field trips, we told the cultivators of our intention to interview them and thus solicited their cooperation, which was readily granted. Every cultivator we approached was willing to grant us interviews.

Most of the data used in this work were collected through field interviews of urban cultivators, government officials and non-cultivators in Accra. In addition, some urban farm labourers were interviewed. However, the data gathered from the labourers were not used in this work because they turned out not to be of any importance to urban agriculture in Accra. For example, urban labourers do not affect cultivation in any significant way because their services were not much used by the cultivators.

I hired four research assistants to help me with the interviews. That means, in all, five people conducted the interviews for this study. It should be mentioned that my research assistants helped in interviewing the cultivators only. It was myself who interviewed the government officials and the non-cultivators involved in this study.

The fieldwork consisted of semi-structured interviews covering the areas, as shown in Map 2, into which the city was divided. Apart from the interviews, I spent two weeks observing the group of cultivators at Dzorwulu.

In addition to the field research, desk research was also used. Some

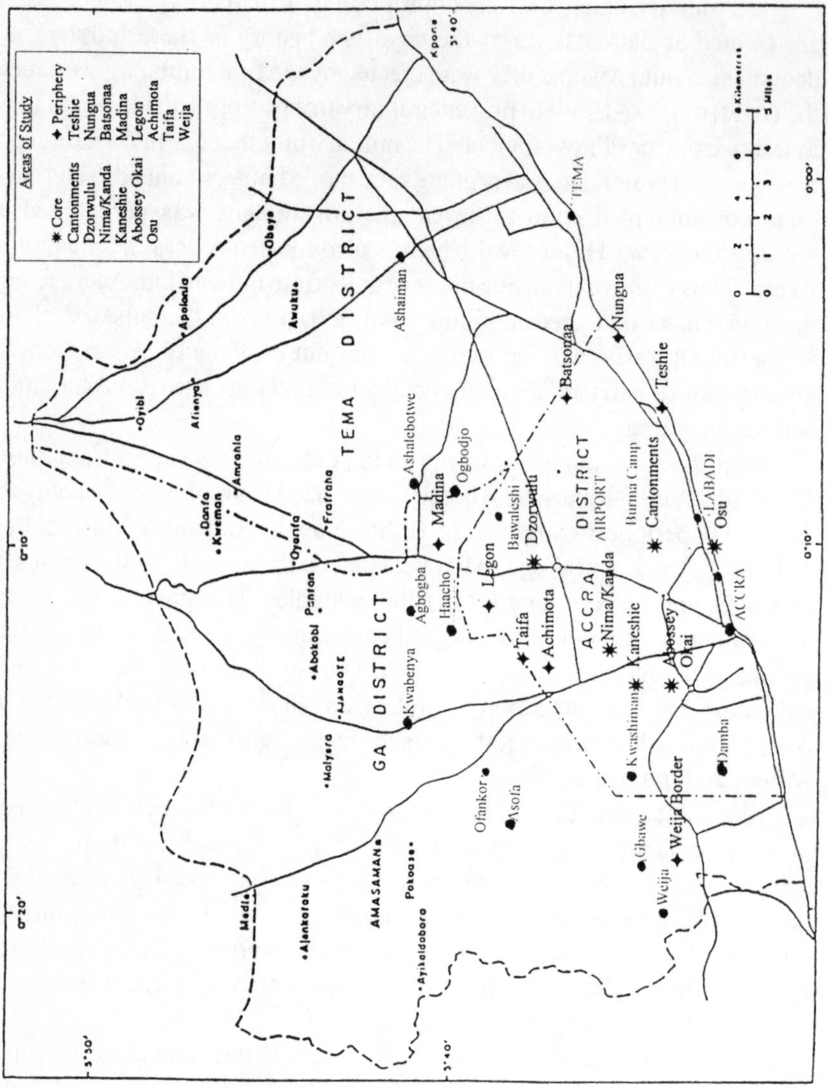

Map 2 Accra: Areas of Study

important data were gathered from existing books and articles. That means, in some circumstances the research findings of other social scientists were relied upon. It should be mentioned that statistics collection in sub-Saharan African countries is not adequate or up to date. Therefore, on a few occasions I had to rely on incoherent data/statistics. Incoherent in the sense that there may be statistics on, let's say, educational level of women in Ghana for 1989 but there may not be statistics for 1990, and again there may be statistics for 1992.

3.3 Sampling

Two hundred urban cultivators were involved in this study. In addition, 40 urbanites who were not engaged in urban agriculture at the time of this study, and 12 government or state officials were interviewed. This means, in all, 252 respondents were involved in the Accra study.

Of the 200 cultivators, 150 were open-space cultivators and 50 were enclosed cultivators. A larger number of open-space cultivators was chosen because at the time of the study, open-space cultivators seemed to form the majority of urban cultivators in Accra, and they produced mostly for sale. Consequently, they were more important for the country. The enclosed cultivators were chosen by purposive and structured sampling. In each of the areas under study, we decided on the number of enclosed cultivators to be interviewed. After that, we decided every tenth house would be involved in the study. We followed this system until we reached our target of the number of people to interview in an area. It should be mentioned that in each of the six places of study in the core area of Accra, three enclosed cultivators were interviewed: this means, a total of 18 enclosed cultivators in this study were farming at the core area. At the periphery, 32 enclosed cultivators (four at each study area) were interviewed.

Since not every enclosed household was involved in cultivation, we made it a policy to try the nearest house to our target house if the target household was not involved in cultivation. For example, if house A was a tenth house, and we found out that urban agriculture was not practised there, we tried the house nearest to A. For the purpose of this work, this method of sampling was the most appropriate because this category of cultivators live and keep gardens within their walled compounds. That means, in many cases, we did not know beforehand which house was cultivating and which house was not. We found this out only after entry into the compounds.

Open-space cultivators normally cultivate in clusters. So after purposely demarcating the areas they were farming, we used unstructured methods to choose our respondents. Thus, we did not follow any special pattern but interviewed any farmer at the clusters willing to talk to us. In each of the eight areas of in the periphery, we interviewed 12 cultivators. This means, at the periphery a total of 96 open-space farmers were interviewed. At each of the six areas in the core, we interviewed nine cultivators. This is a total of 54 open-space cultivators at the core.

I should mention that on a few occasions we relied on snowball sampling to identify the clusters of open-space cultivators. For example, as noted above, after I had demarcated the city into areas, we conducted an initial survey to find out where the cultivators were. When I went to Cantonments, some of the cultivators there mentioned an area at Dzorwulu where urban cultivation takes place. Some of the cultivators at Cantonments even mentioned names of friends who were cultivating at Dzorwulu saying I should tell them that they (at Cantonments) recommended them to me for the interview.

Since their farms are in the open, it was easier locating the farms of open-space cultivators than those of enclosed cultivators. Both types of cultivators were equally willing to grant us interviews.

As earlier mentioned, some non-cultivators or urbanites, who were not involved in cultivation during the time of this research, were interviewed on the streets. For unknown reasons, some urbanites I tried to talk to refused to listen to me. However, this attitude was anticipated so it did not discourage me. The street interviews were done in the mornings, afternoons and in the evenings in order to make sure that I did not interview only a particular group of people. For example, if I had conducted this aspect of the interview only in the afternoons, many formally employed workers would have been left out since they would have been at work. Sampling, as far as non-cultivators are concerned, was unstructured in the sense that any person who passed by was stopped for an interview. It was conducted mostly at various parts in the core area of Accra. However, a few of the non-cultivators were interviewed at the periphery.

Prior to the commencement of the fieldwork, I decided on the types of government institutions to be included in this study. After that, I went to the institutions and explained my mission to the receptionists who directed me to some officials. The usual trend was like this: the first official I met recommended some other official (his/her boss), and this one in turn sent me to some other official. This went on until some of them agreed to be interviewed. At times, I came back to the first official I approached after his or her boss had redirected me to him or her. It was an issue of seeking consent from a superior before granting an interview, that is, if the superior him/herself could not grant the interview. The former indicates that it was more difficult getting the consent of government officials than cultivators to interview. It should be mentioned that after getting a consent from officials and booking an appointment, it was also difficult getting some of them to honour the appointment. On a few occasions, I met up as agreed upon, just to be told that the official did not turn up for work or that he was busy with other assignments. This did not discourage me since I knew the officials had some duties to perform in their respective offices. Making themselves available for an interview was a privilege they granted me. Eventually, all the government officials I fixed interview appointments with were interviewed.

3.4 Sample Characteristics

The age groups and gender distribution of the cultivators involved in this study are shown in Table 3.1.

Table 3.1 Age/Sex of Urban Cultivators

Age Group	Males	Females	Row Total	% Cumulative
15-24	8	0	8	4.0
25-34	25	5	30	19.0
35-44	48	30	78	58.0
45-54	39	19	58	87.0
55-64	14	5	19	96.5
Over 65	5	2	7	100.0
Total	139	61	200	

The table above shows that 4 per cent of the cultivators in this study (all of them males) were between the ages of 15 and 24 years. Thirty cultivators (15 per cent) were between the ages of 25 and 34 years. Out of this number, five are females and the remaining 25 are males. Most of the cultivators (39 per cent) in this study were between the ages of 35 and 44 years followed by the cultivators in the 45 and 54 age group (29 per cent). In the 55-64 year age group there were 19 (9.5 per cent) cultivators, while there were only seven (3.5 per cent) cultivators who were above 65 years of age. In all, there were 139 (69.5 per cent) male cultivators and 61 (30.5 per cent) female cultivators.

The officials interviewed were from the following institutions. Two from the Policy Planning Monitoring Evaluation Department (PPMED) of the Ministry of Agriculture, two officials of the Extension Services also of the Ministry of Agriculture, four officials of Accra Metropolitan Assembly (AMA) [two from the Waste Management Department and two from the main office], two officials of the Environmental Protection Agency (EPA), and two officials of the Parks and Gardens. Apart from Accra Metropolitan Assembly, which is a local institution, all of these are national institutions.

The officials of PPMED were included in the study because of their duty of planning, monitoring and evaluation of the duties of the Ministry of Agriculture. The Extension Services were included in the study because it is their responsibility to 'render services to farmers' (an official's words). It was necessary to include the EPA because of its assignment of protecting the environment. AMA officials were vital for my study because they are in charge of the administration of Accra. The Department of Parks and Gardens control large plots of undeveloped land—land that is constantly encroached upon by urban cultivators, so it was necessary to include them in this study.

As far as urban farmers are concerned, the most important establishment is the Extension Services of the Ministry of Agriculture because they expect assistance in the form of advice on improved methods of farming and the proper use of chemicals from the Agricultural Extension Services. Next in importance is the Department of Parks and Gardens because, as earlier mentioned, this Department controls many plots of land in Accra, and the farmers wish the Department releases some of the Department's plots of land to them for cultivation. The Waste Management Department of AMA is also vital for the farmers because of its production of compost but due to distribution constraints, the compost does not reach most of the cultivators.

3.5 Data Collection Procedures

A full year was spent in the field, meaning the study took place during a dry and wet (growing) season in Ghana. The dry season is normally from November to February, and the wet season starts from March through to October. Spending a year in the field means that my assistants and I were able to study urban farmers during land preparation, planting, and harvesting periods. This means we could conduct the study at the two peak or growing seasons of March/April and September, when Ghanaian farmers were very busy, and the lean season when they were less busy.[3] The units of analysis of this research are urban farmers, government officials, and some non-cultivators in Accra.

Data collection was mostly through qualitative interviewing and on a limited basis through the observation of cultivators. I had an interview guide of 135 questions for the cultivators, 34 questions for government officials, and 20 questions for non-cultivators. The questions were open-ended, so as to give the respondents the opportunity to express themselves as much as they wished. However, the way some of the questions were framed necessitated short and straight-forward answers. For example, a question like 'have you ever lived in a rural area?' necessitated a 'yes' or 'no' answer, and a question like, 'how many people are in your family?' demanded a specific figure.

All open-space cultivators were interviewed in their farms. Eight enclosed cultivators were interviewed in their gardens, and seven were interviewed in their offices. Concerning the remaining 35, the interviews took place in their homes.

It should be re-emphasized that some of the data used in this work are from secondary sources. The sources of such data are clearly acknowledged.

3.6 Data Analysis

In order to be able to pull out frequencies and cross-tabulations from the data collected, I categorized and coded the answers from the respondents. For example, looking at the ages of the cultivators, the answers were grouped into deciles of 15-24, 25-34 years, and so on.

Due to the fact that the answers were grouped and coded, I could use SPSS 6.1.2. for Windows programme to analyze my data. My main intention of using the SPSS programme was to identify the frequencies of the answers.

---

[3] It should be noted that most open-space cultivators have a reliable source of water throughout the year. That means they cultivate all-year round.

3.7 Comments on Methodology

As noted in the previous section, the answers were categorized after, instead of before the interviews. This approach seems to be a departure from common practice, where categorizing is done before interviews are conducted. Consequently, it may be seen as affecting the validity of the data collected. However, my approach did not affect the content of the data collected because the categorization was based on the scale of answers provided by the respondents. It may rather be advantageous because the answers determined the categorizing, meaning the respondents were not compelled to choose within categories.

English, pidgin English and some Ghanaian languages, depending on the educational background of the respondents, were used during the interviews. Thus, respondents who could not express themselves in English were interviewed in a Ghanaian language or pidgin English. It is pertinent for me to mention that most of the respondents who could not express themselves in English, were in the lower socio-economic class category of farmers. It is also pertinent to mention that some of the quotes attributed to open-space (lower class) cultivators are translations from either a Ghanaian language or pidgin English. The translations were done in order to make their expressions readable to non-Ghanaians.

An important problem with the data is related to the small number of cultivators involved in the study. However, this is compensated for by the large number of questions posed to the respondents. Generally, the most important problem encountered is associated with lack of official statistics. In the end, I had to eliminate some aspects of my work due to lack of supporting data.

Having written about the methodology, in the next chapter I turn to a discussion on the emergence of contemporary urban agriculture in sub-Saharan African towns and cities.

4 The Emergence of Contemporary Urban Agriculture

4.1 Introduction

The main purpose of this chapter is to conceptualize the emergence of urban agriculture in a typical, previously colonized country. In addition, an attempt will be made to point out when urban agriculture became an issue of concern in Ghana meaning, i.e. at which point in time urban residents started cultivating in the towns and cities at a high rate. I hypothesize that in a typical modern sub-Saharan African country like Ghana, urban agriculture will become more of an issue for the first time when economic situations are difficult—that is during severe economic recession. I do not intend to give the impression that the rate of urban agriculture fluctuates with economic conditions. After an initial major involvement of many urban residents in urban agriculture, I expect the trend to go up, irrespective of fluctuations in the economy. That is after initial major involvement, even if the economy improves, urban residents will still be involved in urban agriculture. Where necessary, I will draw on experiences from several countries to support my arguments.[1]

Any sort of urban agriculture that existed prior to colonization is ignored despite the fact that 'unlike Western European counterparts, African cities and towns were basically agrarian' (Hull, 1976:388). Archaeological evidence

---

[1] It should be noted that though Africa is not a homogenous place, there are several similarities between countries, especially those colonized by the same countries.

shows that the early cities 'required food storage and this led to the selective domestication of animals and to the regeneration of seed stock' (Jacobs, 1970:47). In Ghana for example, early colonial travellers reported aqua-terra farming systems (UNDP, 1996). As 'late as medieval Europe, farming was an important urban function. How could it have been otherwise without modern transportation?' (Winters, 1983:15). According to Winters,

> In a hot, often humid region such as tropical Africa, the problem of storing food was added to the problem of transporting it; and the fact that urbanism was so independent of trade was one more reason for cities to be self-sufficient in food (Winters, 1983:15).

It is also noted that before colonization, urban agriculture 'played a major role in pre-colonial cities of non-Islamic origin, especially those which were seats of rulers and religious centres' (Winters, 1983:7). West African towns and cities, for example, were surrounded by 'a zone of intensive farming where a large part of the urban population worked each day' (Bowditch, 1819:322). Agriculture played an important role in Eastern and Central African cities also. In these areas the 'quarters of which cities were made up were frequently separated and the space between them used for farming' (Winters, 1989:14). As an observer said of Kampala, 'it was less of a city than an immense garden' (Gutkind, 1963:9). The capital of Loango in Congo was so green that an eighteenth century missionary reportedly remarked that 'a missionary who was a bit nearsighted could have traversed the whole town without seeing a single house' (Balandier, quoted in Winters, 1989:14).

The preceding observations indicate that urban agriculture existed in pre-colonial Africa. However, during colonization it was banned, and in some countries it is still banned, or at best not recognized.

This work ignores pre-colonial urban agriculture because not much is known about how it was practiced. Secondly, during that period almost everybody was a cultivator—urban and rural dwellers alike—so urban agriculture was not any issue of critical examination, at least as far as this work is concerned. The advanced specialization we have today did not exist during the precolonial era. At that time, household units produced almost everything they needed and did not depend much on the products of others. In addition, there was not a large concentration of non-agricultural workers in specific areas. Thirdly, as Winters put it, 'the first cities were completely different from present-day cities' (Winters, 1983:4) so the concept of urbanization as it is used today may be quite different from its use during the pre-colonial era.

Therefore, urban agriculture as it is studied in this work is assumed to be a colonial phenomenon when, it was expected that agriculture be reserved for rural residents. In this regard, urban areas were expected to be residential, commercial, administrative and industrial areas, and rural areas places of food production. The colonizers who controlled African cities had 'concepts of grandeur, precepts of cleanliness and a firm intent to distinguish themselves from the *bush*' (UNDP, 1996:38). Therefore, cultivation of crops and rearing of animals were not permitted in the cities. It should however be mentioned that at the very beginning of colonization some of the colonists themselves encouraged urban agriculture. For example:

> When the Danes entered the West Coast gold trade they caused tracts of land to be cleared in the Accra area and they employed the inhabitants in a form of plantation farming... therefore, a strong foundation was laid in these areas for sound agricultural practices (La Anyane, 1963:9).

The encouragement of urban agriculture at the beginning of colonization happened prior to the opening up of the hinterland. It was therefore a measure of necessity. When the colonists were able to open up the hinterland to ensure the transfer of food to the coasts, they banned urban agriculture.

First and foremost, urban agriculture was not permitted because the colonists thought it would compromise town and city health, that it would distract the so-called natives from working in the emerging formal economy. In addition, since in most cases rural agriculture could provide all the food needed by the towns and cities, farming in the urban areas was not necessary. The only plants that urban residents were permitted to grow were ornamental plants—plants that could beautify the towns and cities.

For the sake of clarity this chapter is divided into historical stages dating from early colonial days to 1996.

4.2 Early Colonial Era (c. 1890-1920): Period of Immobile Labour

At the beginning of colonialism, African labour was not mobile because people worked in their villages, and did not see the economic and social need to move from their villages to other places in the name of work. The economies of most

African societies were based on communism where there was the use of resources for communal well-being, and private profit was not a basic motivating factor (Rau, 1993). The community took care of the needs of individuals, so the individual African did not have to sell his/her labour for survival. In addition, there was an ecological homogeneity so apart from a few activities like mining of some minerals, which necessitated that 'various localities specialized in one industry or the other' (Agbodeka, 1992:15), the different communities looked alike in the form of job provision or job availability.

At the advent of colonialism, the colonists sought to expand the output of cash crops and other activities for export to European countries, but the African immobile labour hindered recruitment of labour to work in the plantations, mines, and other establishments owned by the colonists. African labour was so immobile that 'as early as 1893, a compulsory recruitment law called the Master and Servant Ordinance had been introduced to initially provide ... unskilled workers for the colonial administration in its "opening up" of the country [Ghana]' (Aidoo, 1983:149). Thus, it was eventually through colonialism that Africans were employed to work for the colonists. Therefore, colonialism became 'the instrument for assuring that European home-market demand was met' (Rau, 1993:30).

The few manufacturing, commercial and mining towns which existed or were established, were close to the home villages of the workers employed by the colonists. Consequently, the workers could commute from their homes and villages to their work places in the towns. This is an indication that they lived in the food producing areas and did not have to leave their wives and kin so, they interacted with them on a daily basis. At this period only,

> Two to three per cent of the African population lived in towns, and forty per cent of these were Nigerians whose populations had lived in towns for over a hundred years (Vennetier, 1967a:31).

Many contemporary African cities were created by the colonists. For example,

> Accra, now the capital of Ghana, was initially a number of coastal slave forts but grew rapidly from 1877 when the British moved their administration from Cape Coast. Nairobi in Kenya was established almost accidentally as a railway

> camp in the 1890s in an area lacking in large settlements. The mining towns of Zambia were mostly developed in the late 1920s when rising copper prices made it worthwhile to invest in mines at Chingola, Kitwe, Luanshya, Mufulira and Ndola (Binns, 1994:119).

Since labour was not mobile, only people from the areas of the activity in question offered their labour for sale. For example, only people from the mining areas were employed as miners. Apart from the immobile nature of labour, the hiring of African labour was influenced by existing skill. Due to their mining activities before colonization, people in mining areas had mining skills. It should be mentioned that most of the workers were casual workers, so they could spend a substantial number of days on their farms when they were not working for the colonizers.

At this stage, there was no need for any policies on urban agriculture because it was not an issue of concern to anybody.

4.3 Mid-Colonial Period (1921-1948): Period of Limited Mobile Labour

Owing to increased European activities in sub-Saharan Africa there was an increase in the number of manufacturing, mining and commercial towns during this era. In Ghana for example, between 1921 and 1948, 104 new establishments were created (Ewusi, 1986:4). This subsequently resulted in a faster rate in proletarianization.

Most of the newly emerging towns and cities were further away from the villages of the potential workers, due to the discovery of deposits at areas not known to contain minerals. For example, following survey activities of the Ghana Survey Department, bauxite deposits were found in Asante and Kwahu (see Agbodeka, 1992). Since the newly emerging towns were further away from potential workers, some workers had to move away from their villages to the mining areas. During this period, Africans generally were excluded from urban settlement unless employed, and with the expectation that they were only temporary inhabitants (Rau, 1993). For example, to justify the denial of Africans urban residence, in 1933 an official in Nairobi maintained that 'if they [Africans] go back to the reserves the chances are that their children will grow up decent people. If they continue living in town there is little hope of that' (White, 1983:184).

Some of the new towns emerged from existing villages but as earlier noted, many of them were new settlements. Most of the workers were therefore drafted from elsewhere, they were not indigenous to the emerging towns. As noted by a researcher, '... "free" (i.e. expropriated) labor not only began to migrate, but was also forcibly recruited when needed in the mines and other capitalist production sites' (Aidoo, 1983:149). The workers employed were from other areas, not from the emerging towns, since the casual employment offered in the early towns had accorded many men from those towns the opportunity to work for the colonists. Consequently many of them understood the expectations of the colonists, i.e. they had experience working with the colonists, and they had acquired skills needed in the new towns. By implication, people with previous experience could produce more so they were preferred by the colonists as workers.

It should be mentioned that the Africans were not willing to move away from their local or traditional economy because the cost of living in the emerging urban areas (including the cost of leaving one's wife behind) was higher compared to living in rural areas, and notoriously bad working conditions (see Aidoo, 1983:149). Since the Africans were not willing to migrate, the colonists had to compel indigenous labour to migrate to the emerging modern economy by, among other measures, placing a head tax on adult males. As noted by Ohadike, 'the colonial capitalist state enforced various forms of unpopular taxation by which labour was forced to leave the land to go to colonial mines and plantations for wages to meet its tax obligations' (Ohadike, 1988:117). It is also noted elsewhere that, '... shortages of labour led to the strengthening of mechanisms designed to ensure an adequate labour supply' (Rakodi, 1997:25).

During this era, like the previous one, the majority of workers were male and were compelled to move without their wives and children as these were forbidden by the colonial authorities to join the men in their work areas. According to the colonists, since African children were not living in the towns 'there was no danger of these putatively highly infective individuals being brought into close contact with Europeans' (Freeman, 1991:32).

Fortunately, the distances between the newly emerging towns and the villages of the workers were not very great, so they were able to visit their wives and kin frequently. In addition, migrant workers were initially away from their homes for relatively short periods—two to six months—depending upon the type of job obtained (Rau, 1993). The frequent visits allowed them to take foodstuffs to their towns of work to supplement their food purchases. Conversely, 'they often depended on the food parcels that a wife or visiting relative brought them from farms in their home areas...' (Bryceson, 1987:164). However, most of the food consumed by the workers came from taxes paid in food by rural people (Rau, 1993). Thus, to make food available to the workers at cheap prices, the colonists collected foodstuffs instead of money as head taxes from some peasants. In addition, the colonists engaged in bulk food purchases to provide workers with food in kind (Guyer, 1987:29). In turn, an amount was deducted from the cash wage to reimburse the employer (colonists) for the cost of food supplied. For example, 'many employed Africans resident in Dar es Salaam received a portion of their wage in kind...' (Bryceson, 1987:164).

Urban agriculture was not permitted and there were heavy crackdowns on violators, which made the practice almost nonexistent. In Kenya for example, the colonists rejected the idea of vegetable plots in Nairobi saying 'land values and the difficulties of piped water supply make it impossible to provide separate plots for each dwelling where crops can be grown' (White, Silberman and Anderson, 1948:37).

Also writing about the mid-colonial period, Guyer (1987) notes that 'by the end of the 1930s the urban and the employee populations [of Africans] were still small' (Guyer, 1987:34), and food supplies from the rural areas were adequate to feed them.

4.4 Late Colonial Era (1949-1956): Period of Dynamic Labour

In the late colonial period, workers moved further away from their home villages, especially to the national and regional capitals. It was at this stage that 'urban growth gained momentum in Africa' (Guyer, 1987:35). In Ghana and neighbouring countries, for example, there was much inter-regional migration. Consequently, a large number of the workers in Ghana came from the neighbouring countries. At that time it was estimated that in Ghana, 'the Government staff, excluding Railways, increased between 1945 and 1950 by roughly 60 per cent, the Senior Service from 960 to 1,555, and the Junior Service from 6,770 to 14,100' (Commission of Civil Service, 1951:19). In an attempt to improve upon production, 'both industrial and agricultural capitalism sought to stabilize their forces ... by demanding longer contracts from their workers' (Rau, 1993:40). For the workers, longer contracts meant fewer visits to their home villages. The visits were eventually reduced to major celebrations like Christmas and Easter: festivities the colonists did not consider barbaric.

At the initial stage of this era, labour was still not very mobile as most Africans preferred to cultivate in their villages to pay their head taxes in food. Eventually, the colonists saw head taxes paid in food as a hindrance to free movement of labour, so they rejected it. The rejection was an effort to force more labour out of the traditional economy and into the emerging modern economy. Since the indigenous people were not allowed to pay their taxes in kind, they had to sell their labour to earn money to pay the taxes. Prior to the colonists' demand for longer contracts, different sets of workers could use the same accommodation alternatively. That is, when a group of workers were off duty (when they were in their villages), others used their accommodation. However, the eventual demand for longer contracts necessitated that most of the African workers spent longer periods in the towns. Thus sharing of accommodation in turn became problematic.

There were many negative consequences which resulted from longer contracts and the movement of more labour into towns. However, the most dire one was improper accommodation. In an effort to solve the accommodation problem the authorities set up a commission to suggest solutions. Among other factors the Commission submitted that, 'the provision of adequate housing is one of the most urgent problems facing the Gold Coast Government' (Commission of the Civil Service of the Gold Coast, 1951:15). A Department of Housing was then entrusted to construct subsidized housing estates for lower paid workers.

In addition, the movement of more labour into towns resulted in decreased food production in the rural areas. Furthermore, as visits to kin in rural areas became less frequent the workers did not have adequate food supply from rural-based kin. Also, head taxes paid in food had been stopped, leading to falling food supply to workers. As a result, food became more expensive for the average worker. The problem with food supply to the workers was compounded by the fact that more people were compelled to sell their labour in the mines instead of cultivating the land. In an effort to make food available to the workers at an affordable rate, another Commission was set up to study their conditions. The Commission came to the conclusion that 'civil servants, senior and junior, and daily-rated, were suffering hardship as a result of price increases since 1945-46' (Ministry of Finance, 1958:50).

Increased proletarianization, as witnessed in this era, did not necessarily lead to the expected increase in production. The hardship resulting from the increased movement of peasants and other workers into the towns resulted in alienation of the workers. This in turn reduced relative production. On the basis of the recommendations submitted by the various Commissions, the colonists realized that to improve upon the productivity of the workers some concessions were to be made. Consequently, the living conditions at or near work sites were improved, and in some cases, families were provided for (Rau, 1993) in order to provide worker satisfaction. In addition, cost of living allowance was paid to the workers. Furthermore, the administrators argued for,

> A permanent resident African labour force that will be more efficient, responsible and less threatening than the contingents of single males on short-term contracts. These families could be housed and educated and so become a worthy addition to the cities (Lowder, 1986:92).

In addition to the increase in the African urban population due to, among other factors, family reunions there was an increase in the service sector which necessitated an increased employment of women. As was then noted in Ghana,

> Ghana's towns are growing rapidly, and it will continue to do so, primarily because of the growth of services which will provide much employment for the young boys and girls who are now starting to pour out of the school system (Ministry of Finance, 1958:1).

Apart from the employment opportunities opened to women by the growth in the service sectors, there was also an influx of women into the cities as,

> African businessmen and civil servants, responding to the Africanisation programme ... brought their wives and families to live with them, and the colonial strictures against women and children in the cities were swept away (Freeman, 1991:81).

Towns, therefore, assumed new characteristics from the earlier ones. They were no longer men-biased, nor anti-female and anti-children.

The concessions given by the colonists led to an increase in the general standard of living of urban residents. Food to feed the urban residents came mainly from the adjacent rural areas. Despite the fact that taxes paid in food had been stopped, after the initial stage of hardship, food supply for the workers increased at affordable prices. This was due to the institutionalization of a wage-food price relationship. Thus, the prices of food were deliberately kept low by state marketing boards so that workers could afford them. The urban population was not very large, and the few motorable roads, all leading to the towns and cities and other places of economic importance were very good. Consequently, it was easy to cart food from rural to urban areas. However, it should be noted that some problems still existed with food supply since,

> Food production is not keeping pace with consumption: this is clear from the increase in food imports which have been growing by 7 per cent per annum over the past five years (Ministry of Finance, 1958:2).

In British colonies, by the end of this era limited and controlled urban agriculture was permitted for the benefit of the colonists. This is supported by the following observation in Nairobi,

> In the area encompassed by Ngara Road, Market Road and Government Road lay the Nairobi swamp, at first thickly covered with papyrus but later on cleared for vegetable gardens which were well kept from the beginning... some parts of the park might well be set aside for use as allotments for vegetable growing, the additional area of ground being gained by a slightly closer net density of housing where the amenity of allotments is available (Thornton-White, Silberman and Anderson, 1948:66).

Despite the above-mentioned concession to cultivation in towns and cities, generally the colonial administration 'rejected the idea of servicing vegetable plots in the city' (Freeman, 1991:38) and after independence the new indigenous administration continued with the tradition.

4.5 Nkrumah's Era (1957-1965)

After independence in 1957, the administration of Ghana was jointly in the hands of an educated elite and an economic (business) elite. Most of the members of the educated elite were foreign trained and had a taste and lifestyle different from that of the average Ghanaian. The business elite too had associated themselves with the colonial masters in the towns and had also acquired a craving for foreign food and lifestyle. These elite groups did not support urban agriculture for the same reasons the colonists used to reject it (see chapter 5). Officials of the various Town/City Councils were instructed to destroy growing crops, and animals found outside premises were confiscated. In some cases, cultivators were dragged to courts allegedly for compromising city health.[2]

However, as indicated earlier, some colonists and other expatriates hired some Africans to cultivate back- and front-yard gardens for the purpose of growing vegetables. Consequently, 'some of us who served as garden boys for the expatriates were permitted to cultivate vegetables, only vegetables, to feed our masters' (own interview with an urban cultivator, 1995).

During the earlier stage of this era there was an adequate food supply at affordable prices to the urban population. Among other factors, this was as a result of an increase in the efficiency of the indigenous market system. Toward the end of colonization, the government decreased its involvement in distribution, leaving food distribution mainly in the hands of self-employed women who could go deeper into the hinterland to purchase and cart foodstuffs to urban areas. However, after political independence the new government increased its involvement in food distribution in a different form. The government continued to buy food in bulk to sell to the workers. Through numerous food distribution shops it also sold to the general urban population, thus expanding its operation in this sphere.[3]

Increased government involvement in food distribution had a negative consequence in Ghana. By 1960, due to the increased government intervention (cf Marketing Boards) the distribution system had become less effective. Lawson (1967) noted the growing animosity between the independent government and the redoubtable female traders who ensured the consumer goods market, including the food market. The market control stance taken by the government led to further increases in the prices of food, owing to the inefficiency of the Marketing Boards. They could not reach the peasants in the hinterland to purchase foodstuffs—in turn, distribution was affected because they did not reach most of the consumers. The most important reason was that the distributing depots of the State Marketing Board were not very accessible to most consumers. As earlier noted, the female traders on the other hand, went

[2] Town Council Officials were responsible for keeping the towns and cities in healthy conditions.

[3] The government decided to control the market.

to every remote area to purchase foodstuffs and, in selling, they sold at the doorsteps of consumers. The major consequence of the government's intervention in food distribution can be summed up in a reporter's observation that, in 1959 nutritionally identical diets cost almost twice as much in Accra than in London (Lawson, 1963). Around the same time (between 1952 and 1967), the value of the minimum wage fell by 30 per cent (Knight, 1972), thus exacerbating the plight of the average worker.

The above-mentioned developments indicate that Ghanaian female traders have become a major phenomenon in Ghanaian cities, especially in the area of food distribution. The degree of their involvement in distribution strongly determines the prices of commodities. The increased number of them having easy access into commodity distribution leads to a rise in the supply of commodities. All things being equal, this lowers the prices of commodities, thus increasing the purchasing power of consumers/workers. On the other hand, if the female traders are hindered from distribution of commodities to consumers, supply decreases below demand. This increases prices, and lowers the purchasing power of consumers/workers. It is therefore not surprising that successive Ghanaian governments prefer female traders in the distribution sector.

Some writers like Afful and Steel (1978) and Robertson (1984) have interpreted some measures by Ghanaian governments as repressive measures directed at women traders. For example, Robertson (1984) notes, 'Gender identity is increasingly being used by the governments of Ghana in an ideology which objectifies women traders into a class which can be blamed and persecuted for causing the enormous economic problems' (Robertson, 1984:243). After making this statement she went on to describe measures that were supposedly used to repress Ghanaian market women. The measures mentioned include the expulsion of alien traders in 1970 by the Busia government.[4] Other so-called repressive measures mentioned are, 'the Acheampong government imposed more stringent price controls (first begun by Nkrumah);' 'during the first Rawlings government... soldiers bulldozed Makola No.1 market and reduced it to a pile of rubble' (Robertson, 1984:244). She also notes that less than one third of the revenue collected from the markets was spent on the markets.

It is wrong for Robertson (1984) to interpret the above-mentioned measures as directed against market women. No Ghanaian government has ever come out with a policy that named or singled out market women as

[4] I don't see how this is a repression against Ghanaian women traders.

targets. Acheampong's stringent control measures affected traders or distributors of both sexes, and the bulldozing of Makola No. 1 market was part of a nation-wide campaign against hoarding. It was not a measure against women: many shops owned by men suffered the same fate. That most of the revenues collected from a source are not spent on that source is not peculiar to markets. For example, Ghana's mineral-rich areas, income generating areas, are starved of infrastructure.

Various Ghanaian governments had declared *war* against trade malpractices especially hoarding which eventually created artificial shortages. Robertson (1984) herself gives an impression that hoarding had been an economic problem in Ghana. For example, she quotes a female subject in her studies, 'In our day hoarding of commodities was unknown. Now, women are selling a packet of sugar for almost ten shillings, not because the supply is bad, but because they hoarded it. This is close to highway robbery! The traders forget the poor...' (Robertson, 1984:236).[5]

Many African urban residents—especially those in the middle/upper class—did not cultivate because agriculture in general lost prestige in the eyes of the newly educated and salaried workers. For them, agriculture was the job of the poor and the uneducated. Farmers, apart from cocoa farmers, were in a lower class and those who saw themselves in higher classes did not want to associate themselves with farming. In addition, farming was (and is today in many sub-Saharan African countries including Ghana) tedious due to the use of crude implements, therefore anybody who could avoid farming did so. It will be seen later that even today, of the Ghanaians involved in farming, those of higher social status (rural or urban) hire labourers to do the most tedious aspects of cultivation like land clearing and tree stumping.

Another reason for many African urban residents not to engage in urban agriculture was that prices of staple and non-nutritional food like cassava, which were consumed in large quantities, were still low. The price of non-nutritional food was low because 'the production of plantains and of starchy ground provisions has kept pace with population growth' (Ministry of Finance, 1958:2). In addition, and most importantly, due to the low salary of the ordinary worker most of them had to work longer hours, resulting in little time to cultivate crops.

Aforementioned, non-nutritional food production kept pace with population growth. However, it is documented that by 1957, general food supply in Ghana

[5] By implication Robertson denies hoarding as the cause of high prices—she blames high prices on a 'long chain of distribution' (Robertson, 1984:245).

was not adequate. For example, 'in 1957 Ghana imported food valued at 17,400,000 pounds' (Ministry of Finance, 1958:2). This large importation of food presented a great challenge to improve the productivity of food farming in Ghana.

Confronting the challenge, the government initiated a policy on agriculture. This entailed their active involvement in food production. Government institutions like the Workers' Brigade, the State Farms Corporation and the Young Farmers' League were established to produce food. Government intervention, especially its subsidy on agricultural materials, played a role in the initial increase in food and animal production between 1963 and 1966. See Tables 4.1 and 4.2.

In line with existing procedure, the government turned solely to rural agriculture to boost food supply. Thus various rural agricultural projects were implemented. However, by the end of this era, food production had gone down. For example, according to FAO figures, between 1966 and 1968 maize production dropped by 101,000 metric tons. And within the same period, cassava dropped by 107,000 metric tons (see FAO, 1987:61 and 117).

The above observation indicates fluctuating food production, and in the end the Ghanaian government's involvement through the Workers' Brigade and State Farms Corporation failed.

During this period no effort was made to encourage urban food production. However, through my interviews with urban cultivators I found out that by 1965 some urban residents had begun taking the risks, and cultivated non-vegetable crops like cassava, plantains and maize in the towns and cities. Some of these crops were however destroyed because, as thought by officials, they destroyed the beauty of the cities.

Table 4.1 Domestic Livestock Production 1963-66 (in metric tons)

Livestock	1963	1964	1965	1966
Cattle	476,600	504,556	511,242	527,596
Goats	229,945	319,117	380,477	411,998
Pigs	5,435	73,645	50,903	80,511
Poultry	168,812	811,385	347,773	941,335
Sheep	236,557	332,774	354,677	486,292

Source: Central Bureau of Statistics. Ghana Economic Survey 1966.

Table 4.2 Domestic Food Crop Production 1963-66 (in metric tons)

Crop	1963	1964	1965	1966
All Cereals	392,000	400,000	389,000	647,000
Cassava	1,194,000	1,229,000	1,250,000	1,250,000
Maize	183,000	173,000	180,000	180,000

Sources: 1. FAO, 1987. 1948-1985 World Crop and Livestock Statistics (All Cereals).
2. FAO, 1968. Production Yearbook 1967 (Cassava and Maize).

4.6 NLC/Busia's Era (1966-1972)

The first part of this era was from 1966 to 1968, when the affairs of Ghana were in the hands of a small number of military officers who formed the ruling National Liberation Council (NLC).

There was a strong rising trend in the manufacturing industry due to improved allocation of import licences and the restoration of business confidence (Central Bureau of Statistics, 1970:67). This in turn led to a further increase in proletarianization.

The NLC government and later its successor, the Busia government, disengaged the State from agricultural production. For example, the number of government farms was reduced from 105 in 1966 to 62 in 1967, and to 33 by 1971 (Kraus, 1986:120). The government remained involved in food distribution but only in competition with private distributors. The Food Distribution Corporation was to buy food in bulk, store and release it on to the market at appropriate periods to check food shortages especially during the lean season. Like in the previous era, the government's involvement in food distribution was a failure, as it could not reach the majority of the consumers.

However, by 1969 private distributors made the retail trade so competitive that urban prices for food were below retail prices in rural centres (Reusse and Lawson, 1969). All things being equal, this made food more affordable to urban residents. That was not necessarily the case because by 1970 urban wage labourers in Accra were 'considerably worse off than they had been in 1939' (Sandbrook, 1977:416).

The second part of this era was the Busia era, which lasted from August 31, 1970 to January 13, 1972. During this era a greater emphasis was put on rural agriculture. Subsequently, there were many rural projects to boost food production. Generally, urban agriculture was not encouraged because Busia's government was hopeful of satisfying the population's food need from rural agriculture.

There was no policy or any event to encourage urban agriculture during this era.

4.7 Acheampong/Akuffo's Era (1972-1979)

In January 1972, Busia's democratically elected administration was deposed by a group of military officers. The constitution of the country was suspended, and the soldiers formed the National Redemption Council (NRC) to rule

Ghana.

This era was generally an era of harsh economic conditions brought about by a combination of factors. The immediate factor was the devaluation of the Ghanaian cedi by the previous administration. The remote causes were a general drought at the later part of the previous era, and the servicing of a huge foreign debt incurred mostly during the years following Ghana's achievement of political independence.

A major consequence of the above is that food supply to the population, especially the urban population, again became more of a national issue. In the first place, the economy of the country was at its lowest, and almost every essential commodity was in short supply. The farmers did not have easy access to basic farm equipment like machetes, due to shortages and exorbitant prices. Owing to lack of spare parts and deteriorated roads the public and private transportation systems broke down, thus compounding the problem of carting the few available foodstuffs to the cities. Moreover, the devaluation of the Ghanaian currency by the previous administration had forced up the prices of imported commodities and subsequently local commodities.

A major event exacerbated the food problem. On ascension to power, the populist government of the National Redemption Council attacked previous governments for contracting loans not beneficial to the economy. In a speech to the nation on February 5, 1972, the leader of the NRC (Col. I.K. Acheampong) noted that,

> ... the reports of several commissions of enquiry and other investigations clearly establish that some of these contracts are tainted and vitiated by corruption and other forms of illegality. In some cases there has been a fundamental breach of contract on the part of the contractors. A substantial number of the projects financed by the suppliers' credits were not preceded by any feasibility studies establishing their viability. The prices quoted in respect of these projects were inflated, and the repayment terms did not admit of the projects generating sufficient resources to amortize the debts. While..., a substantial part of the blame equally attaches to the governments of creditor countries. (Acheampong, 1972:4).

The leader of the NRC noted that negotiations by previous Ghanaian governments for long-term debt relief ended in abject failure. As he put it,

> The creditor countries have insisted on their pound of flesh. They have persistently refused to view the solution of our debt problem in terms of a developmental perspective (Acheampong, 1972:4).

On this note he went on to declare that the NRC could not accept any of the debt settlements concluded with creditor countries since February 24, 1966. For the international community the most shocking announcement came when it was made clear that his government,

> Repudiates all contracts which are vitiated by corruption, fraud or other illegality. All debts and other obligations arising under such contracts are cancelled with effect from today (Acheampong, 1972:6).

As a first step, the government cancelled the contracts of four foreign companies with immediate effect.[6]

Not only loans not beneficial to the economy were repudiated. Also debts and other obligations arising from contracts where there had allegedly been a fundamental breach of agreement on the part of indigenous contractors were repudiated. And it was not only contracts tainted with corruption that were repudiated but also those which were not technically and economically viable. That means even in cases where Ghanaian contractors had misappropriated borrowed money, the government was not willing to pay. In addition, the NRC government would not pay for any ongoing project deemed unnecessary. For example, if a foreign company had accepted to implement a project in Ghana, the new government would honour the financial aspect involved only if it deemed the project viable. The onus of viability then rested on foreign companies. Therefore, it was not enough for a foreign company to bid and win a contract in Ghana. The government's and indeed Ghanaians' slogan as far as debt payments were concerned was 'we shall not pay.'

The international community responded with a boycott of credit and other forms of aid to Ghana. For example, food aid was seriously curtailed, and this was a time when due to drought and consequently poor harvest of the previous years, the country relied on some sort of food aid from the international community. To counteract the boycott action taken against Ghana, the

[6] The total value of the cancelled contract was put at US$94.4 million (in 1972 currency).

government launched Operation Feed Yourself (OFY). According to a commentator, 'the OFY was the most ambitious programme ever launched in the country to respond to the [Ghana's] food problem' (Hansen, 1989:203). The main purpose of this official programme was to make Ghana self-reliant in food by encouraging Ghanaians (urban as well as rural) to grow their own food. Urban agriculture was not specifically mentioned in the operation. However, it can be concluded that this was the period when urban agriculture became an issue in Ghana—the period when many people from all the various status groups entered into urban farming. During this era, as indicated by the hypothesis at the beginning of this chapter, economic hardship compelled many urban residents to enter into urban agriculture (see Table 5.1, chapter 5).

In an effort to make OFY successful, the government subsidized the prices of agricultural implements including seeds, fertilizers and chemicals. This in turn led to a spilling over effect of OFY on urban agriculture. The Ministry of Agriculture opened Agro-chemical shops all over Ghanaian towns and cities. Consequently, urban farmers could purchase seeds, seedlings, fertilizers, and other agricultural inputs at subsidized prices. This era was the most important period in the history of the country, as far as urban agriculture is concerned because gardens sprang up all over urban Ghana.

Though OFY was not launched specifically for urban agriculture, it gave urban residents the opportunity to farm without any fear of their crops and animals being destroyed by government officials. During the time of food shortages, and especially during the time the government encouraged Ghanaians to increase their food production, it was unlikely that a government official would have any grounds to destroy growing crops. The incidence of encroachment of public land in urban areas increased with impunity, and until a piece of land was due for development it was not reclaimed by the government. The government of Ghana came out with some by-laws regulating the types and number of crops and animals that could be cultivated or raised in urban Ghana. Consequently, it could be said that the government inadvertently enacted some policies regulating urban agriculture. The by-laws are discussed in chapter 5.

4.8 Rawlings' Era (1979-1998)[7]

There was Limann's 15 month rule between September 1979 and December 1981. This administration purported to increase rural agricultural production. However, by the end of the regime rural food production had not improved. Limann's regime did not affect urban agriculture in any significant way. Urban agriculture remained as it had been in the preceding era. Therefore, there is nothing to write on urban agriculture during Limann's short period of governance.

In June 1979 the Rawlings administration took over the governance of Ghana. The government attempted, as far as food is concerned, to intervene at market level. Concern for security of urban commodity supply compelled the government to take drastic measures to ensure adequate supply. Goods including foodstuffs were seized from suspected hoarders and in turn sold to the public at government recommended prices, which were considerably lower than the prices the traders had intended to sell them at. Between June 1979 and September 1979 of Rawlings' era (first part), the government's policies and actions were of a short-term nature and did not have any significant impact on urban agriculture.

However, after the brief interlude (Limann's governance), Rawlings' administration moved from short-term to long-term agricultural policies. For example,

> In conformity with the regime's bias towards an open market economy a phased withdrawal of subsidies on insecticides, spraying machines, fertilizers and other vital agricultural inputs as well as reliance on private as opposed to public distribution mechanisms have been instituted (Gyimah-Boadi, 1989:237).

The long-term agricultural policies of Rawlings' government had some consequences on urban agriculture. For example, the withdrawal of subsidies on agricultural inputs increased the production costs of urban farmers, which resulted from the increase in prices of equipment they used. Urban farmers had two main ways of adjusting to this shock. They could decrease or stop using farm inputs like fertilizers. Alternatively, they could increase the prices of their

---

[7] This era will end in year 2001.

farm products to reflect the increased cost of production. The option chosen was partly determined by the socio-economic position of the individual cultivators, and partly by their marketing goals. That is, an option was chosen based on whether the farmer produced for sale or for consumption.

The first option is likely to be employed by all types of urban farmers, i.e. whether they farm for sale or for home consumption. However, on the basis of the economic positions of farmers, this option is more likely to be employed by farmers from the lower socio-economic status because they have less access to resources. I discussed the removal of agricultural subsidies with urban farmers in Accra. The words of one of the lower class farmers who cultivates for sale sums up their concern. He said,

> The removal of subsidies affected the prices of our inputs. Prices rose so high that for poor cultivators like us we had to reduce our purchases of important inputs like fertilizers and insecticides. In some cases albeit for short periods, those not financially prepared for the increment had to abandon the use of some chemicals because they could not afford them. At the initial stage, demand for agricultural inputs went down but that did not reduce prices, it rather compounded them. You see, when demand went down importation of these items went down. Then, after the shock and we were prepared to purchase more inputs there wasn't enough in the market so demand rose higher than supply. This shot prices up. It took sometime for prices to level up.

Those who farm for home consumption, i.e. middle/upper class farmers, had a different reaction to the removal of agricultural subsidies. As one of them put it,

> We needed the materials in order to have better harvests so the increase in prices did not affect our demand. You see, when you need something, something that would eventually pay itself you buy it irrespective of its cost.

Those who cultivate for sale, that is open-space farmers (see chapter 7), felt the impact of the removal of agricultural subsidies more than enclosed farmers. In the first place, they were poorer, so the increase in prices of inputs as a result of the removal of agricultural subsidies was too heavy for them to contain. Secondly, they used more inputs like fertilizer than enclosed farmers. Consequently, they spent more money on such inputs. An increase in prices therefore affected them more.

It is expected that a decrease in the application of fertilizers would be made on a rational economic basis. That is, a decrease would be made without subsequently affecting the amount of crops and animals produced because a constant production level might be achieved through the use of improved production methods. Both open-space and enclosed cultivators in Accra noted that the increase in the prices of agricultural inputs encouraged them to learn more in order to improve their production. According to some of them, that was the first time they started approaching agricultural experts, albeit privately, to learn more about improved methods of cultivation. They followed all known agricultural practices in order to avoid waste.

The second option is opened to those cultivating for sale. By increasing the prices of their products they attempt to maintain their profit margin in spite of an increase in the prices of farm inputs. A successful implementation of this option is dependent on consumers' willingness to pay more for the same product. This in turn is influenced by several economic factors. The first factor is the dependency on or dispensability of the commodity: that is, the elasticity or inelasticity of the commodity. If consumers can afford to do without the commodity then they are less willing to pay more for it. The second factor is the substitutability of the commodity in question. Consumers may shift to an alternative product if the original commodity is substitutable. The final factor is, consumers would not be willing to pay higher prices for a commodity if there were cheaper sources of supply.[8]

Food in general is a necessity. However, since there are many types of foodstuffs one type of food may be chosen instead of another. This implies that foodstuffs as individual items may not be seen as indispensable. However, the crops cultivated for sale in urban areas are mostly exotic crops like cabbage, carrots and cauliflower. These are foodstuffs many consumers would not want to do without. Considering the nutritional value of cabbage and other exotic crops, it can be concluded that the consumers do not have any alternatives. The implication is that consumers would be willing to pay more for the same quantity of exotic crops.

An analysis of my data revealed that the majority of urban cultivators in Accra cultivating for sale combined both options to cope with the increase in

---

[8] I have intentionally desisted from mentioning the use of force as a means of acquiring a commodity.

the prices of agricultural inputs. They increased the prices of their harvested crops while exercising austerity in the use of fertilizers, chemicals and other agricultural inputs.

Urban agriculture under Rawlings' later regime has been left to the whims of market forces. Thus, no conscious intervention was made by the government.

Following the labour-surplus theory (see chapter 2) it could be expected that apart from 1972 (the year Operation Feed Yourself was implemented), many urban residents became urban farmers in 1982 and the subsequent years. In 1982 the government negotiated with the International Monetary Fund (IMF) for a loan which culminated in the so-called Structural Adjustment Programme. The features of this programme are,

> ... a reliance on market mechanisms, the promotion of exports, reduction in the size and functions of the civil service, privatization, the elimination of marketing boards, and currency devaluation (Rothchild, 1991:3).

Thus there was an economic restructuring to 'find solutions to those economic conditions that have perpetuated the economic crisis of the recent past' (Oni, 1987:208). Incidentally, solutions to the crisis, especially privatization and the rationalization of public enterprises, led to the retrenchment of the public sector during which several workers lost their jobs.

The redundant workers were encouraged and given incentives to move to the countryside to engage in agriculture. However, instead of relocating to rural areas, the majority of them stayed in Accra and other towns and cities and worked in informal sector activities including agriculture. Most of those who went to the rural areas came back to the urban areas within a year. This compelled the government through the Re-deployment Secretariat of the National Mobilisation Programme to make an announcement in 1991 asking,

> Heads of government departments and institutions who have re-engaged persons laid off from the public sector to take immediate steps to correct the anomaly in their own interest (West Africa, 1991:3121).

The most important contribution of the Rawlings' regime to urban agriculture was the retrenchment that led to an increase in the number of people involved in the practice.[9] In addition, President Rawlings directly stopped the Department of Parks and Gardens from evicting cultivators from its land (see chapter 5).

4.9 Summary

Available information indicates that urban agriculture existed in pre-colonial sub-Saharan Africa. However, I have ignored pre-colonial urban agriculture in this work. I consider urban agriculture a colonial phenomenon, when countries were strictly divided into rural and urban.

During the early colonial era (1890-1920), African labour was immobile because Africans were not interested in moving from their kin to work for the colonist. When eventually some of the people were compelled to work for the colonists they (Africans) lived among their own kin because the first European investments were located near workers' living places. Due to the fact that they lived among their kin, the workers got their food supply from kin members. Thus, those who spent some of their time working for European capitalists and for that reason could not produce their own food, got food supplied by kin members.

There was an increase in mining and other such work during the mid-colonial period of 1921-1948. Still the indigenous people were not interested in working for the colonists, so the latter introduced taxation, thereby compelling the former to sell their labour. Some of the work places were far away from the workers so they had to move further away from their kin. However, they were not too far away, and since they were mostly casual workers they could visit their kin frequently. When going back to the areas where they worked, the workers took with them foodstuffs from their kin. Consequently, they were not short of food supplies. In this era, like the previous one, urban agriculture was forbidden.

The late colonial era (1949-1956) saw an increase in the number of workers, and most of them moved further away to national and regional capitals. The workers were given longer contracts so they stayed in the towns for longer periods before visiting their kin. And their visits to their home areas were limited to shorter periods, so they could not cultivate when they went to the rural areas. The longer contracts extended to the workers created

[9] The retrenchment compelled many urban residents to get involved in urban agriculture.

accommodation problems for them in the towns. In addition, longer contracts meant decreased food production and also a decrease in food supply or food gifts from kin. All this led to worker alienation, and consequently lower production in the capitalist (colonist) economy.

It should be mentioned that the introduction of taxation could not compel many Africans to sell their labour. This is because people could pay taxes in food: they preferred to till their land and use some of their harvests to pay taxes. To compel the Africans to sell their labour, the colonists abolished taxes paid in food. During this era limited and controlled urban agriculture was permitted.

Nkrumah's era is dated from 1957 to 1965. This spans the time of political independence from the colonists and the time the first indigenous government was removed from power. At independence, the administration of the country came into the hands of a foreign-trained educated and economic (business) elite. Like the colonial administration these elites discouraged agriculture in Ghanaian towns and cities.

Just before independence, the government limited its involvement in food distribution. Food distribution was mainly performed by an efficient system of female distributors, making food available to urban residents at lower prices. However, after independence the government reverted into food distribution, and also entered into food production. The government apparatus was so inefficient in the areas of food production, and distribution that general food production, and distribution went down. Consequently, food prices in Ghanaian towns and cities went up. Despite this, many urban workers did not farm. In the first place urban agriculture was banned, and secondly due to their low salaries most of the workers had to work longer hours, therefore having no extra time for urban agriculture. During this era Ghanaian elites regarded agriculture as less prestigious.

During the NLC/Busia era of 1966 to 1972, industrialization and proletarianization increased. The government pulled out of food production and limited its involvement in distribution. Once again food production and distribution became mostly private ventures. The government laid emphasis on rural agriculture and had a lukewarm attitude towards urban agriculture.

The Acheampong/Akkufo era of 1972 to 1979 was an era of extremely harsh economic conditions. A threatening insufficient food supply compelled the government to make food supply its main concern. In an effort to increase food production and supply, the government introduced a programme called Operation Feed Yourself. The main aim of the programme was to encourage Ghanaians (both urban and rural) to cultivate as much food as possible. The government subsidized the prices of agricultural inputs. This was the era in which many urban residents, irrespective of their social status, cultivated in urban areas for the first time. The government introduced by-laws that regulated urban cultivation, meaning by implication, urban agriculture was approved.

During the Rawlings era from 1979 to 1997, the government removed subsidies on agricultural inputs. This increased production costs of urban farmers and it compelled them to exercise austerity in their use of agricultural inputs like fertilizers. Due to an implementation of a Structural Adjustment Programme many workers in the public sector were retrenched. The retrenched workers remained in urban areas and got involved in urban agriculture, thus increasing the number of people involved in the practice. At one stage, the President of the country personally intervened on behalf of some urban farmers, who were threatened by the Department of Parks and Gardens with eviction from the land they were cultivating.

The above indicates an increasing condoning of urban agriculture in Ghana. In the next chapter I will look at the factors that changed government officials' attitude toward urban agriculture, making them more positive to the practice.

5 Official Attitudes Toward Urban Agriculture

5.1 Introduction

It may take many more years before sub-Saharan African political leaders and government officials formally approve of agriculture in African cities or incorporate urban agriculture into urban planning. This is because some officials strongly oppose urban farming. For example, in reply to a question an official of Ghana's Parks and Gardens said, 'we've taken so many steps to minimize it [urban agriculture] but unless the government comes in [politicians intervene] we can do little.' This official insisted that urban agriculture should not be allowed in areas of Accra which attract tourists. A similar attack on urban agriculture was openly made in Zimbabwe by the Minister of Mines. At a rally in 1982, he said:

> If people want to go into full scale farming, they must apply to Government for proper resettlement. They produce nothing but cause a lot of siltation of our sources of water. They must go before they do more harm than good. Zimbabweans must know that urban areas and farming areas are different. They must not do as they please. They must follow the regulations and adhere to them. City Councils must act now and stop this (Mbiba, 1995:92).

Some officials are so negative to the practice of urban agriculture that they behave irresponsibly in efforts to discourage it. An example from Zimbabwe will suffice. After slashing crops in Harare, an official of *the council* said:

> ... now that maize and vegetables have been slashed, it is

> essential that grass cutting be speeded up. Otherwise residents would not see the relevance of slashing their maize whilst grass which is the breeding home for mosquitoes is left to grow (Mbiba, 1995:95).

This slashing of crops is irresponsible, more so, when weeds are not slashed. My point is, the authorities should have used their resources to clear the city of weeds in stead of destroying food crops.

Despite the fact that some government officials oppose urban agriculture, there is evidence to suggest that government officials or bureaucrats in many African countries increasingly are adopting a more positive stance toward the practice (see Mbiba, 1995; Mougeot, 1994; Maxwell and Zziwa, 1992; Mlozi, Lupanga and Mvena, 1992). This is important for the farmers, especially lower-class farmers, because their success in the practice is highly depended upon the attitude of government officials toward the practice. If officials frustrate the farmers, they will not be able to farm in peace and consequently will not be able to produce enough to feed themselves and the urban population in general.

The change in official attitude toward urban agriculture is due to the influences of both economic and socio-political factors, which are studied in this chapter. Consequently, in this chapter, I will examine the economic and socio-political factors that brought about changes in official attitude towards urban agriculture. I hypothesize that:

1. All things being equal, government officials in Ghana and other sub-Saharan African countries condone urban agriculture if the country's economic situation is bad or is getting worse. That is, if the cost of living is very high. This is because during difficult times officials see urban agriculture as one of the means to alleviate hardships, especially hardship related to food shortages and unemployment. If urban residents are not allowed to subsidize their food purchases with food they cultivate, and if urban residents are not allowed to grow food to sell in order to make money (thus employment), they may become more alienated from the government in power. Thus, economic hardship, denial of urban residents to farm in order to feed themselves, and alienation from the government may compel workers or urban residents in general to agitate for change of government.

When officials are to make decisions on urban agriculture they will be influenced by the future consequences. The consideration of future consequences in decision making is described as the theory of anticipatory choice, which stipulates that individuals or organizations act on the basis of some conception of the future consequences of present action for preferences currently held (March and Shapira, 1992:274). As mentioned above, if the economic situation looks bleak at present and in the future, then officials will anticipate food scarcity and consequently condone urban agriculture.

2. Officials would be more willing to allow urban agriculture if many urban residents approved of it. Here, my reference point is that government officials do not want to antagonize urban residents, so they will condone an activity like urban agriculture which may be prohibited but unharmful.

3. The higher the social class or status of urban farmers, the more willing officials are to condone the practice. That is, the higher the status of urban farmers, the more prestigious the practice, and consequently the less likely it is to be prohibited. Similarly, when many government officials themselves are involved in urban agriculture they are less likely to discourage it.

Before I examine the change in official attitude toward urban agriculture, I will generally examine why some officials hold (or held) negative notions about the practice.

According to several historical materials, African cities had a bucolic character. Agriculture in general,

> Was so deep-set that it was carried over into the colonial and postcolonial cities, though it has fit poorly with the preference of colonial rulers and contemporary planners for geometrical order (Winters, 1982:145).

The above indicates that urban agriculture existed during the precolonial period. As noted in the main introduction to this work, the precarious food situation in many sub-Saharan African countries necessitates that urban agriculture in the region should be a potential area for encouragement and development. The need to encourage urban agriculture is more apparent when one considers the fact that the very survival of some urban residents depends on it. The question, therefore, is why is urban agriculture still 'largely unrecognized and unassisted if not outlawed or harassed [in some countries], even in years of food shortage?' (Mougeot, 1993:2).

5.2 Reasons for Official Negative Attitude Toward Urban Agriculture

According to existing literature, urban agriculture is not officially encouraged because of the supposed hazards associated with it. Farming in cities is thought to be unhealthy.

Many officials agree that the use of biocides for pest and disease control in crop production plays an important role in ensuring increasing food supply for the growing population. Biocides check pests and diseases and, consequently, help enormously to reduce food losses. However, 'questions have been raised concerning their effects on human health and the environment' (Goodland, Watson and Ledec, 1984:137). For example, 'the use of a lot of biocides in urban agriculture has been linked to the bioaccumulation of heavy metals and synthetic organic compounds in aquatic life, particularly fish' (Chimhowu and Gumbo, 1993).

A biocide like DDT has been linked to the death of birds (Hardin, 1972). In 1987 it was estimated that 'approximately 10,000 people die each year in developing countries and about 400,000 suffer acutely from pesticide poisoning' (World Commission on Environment and Development, 1987:40). Similarly, the World Resources Institute (1992) notes that runoff of fertilizers, herbicides and pesticides into urban rivers or streams is a significant source of water pollution. In addition, the use of chemicals in food production is also thought to contaminate soils and crops. And the World Resources Institute warns that,

> The threat pesticides pose to human health is particularly potent in the developing world, where most serious exposure occurs. Indeed, pesticide poisoning is disconcertingly common in developing nations, representing a major occupational hazard for farmers and their families (World Resources Institute, 1994:114).

In Ghana, the fear of contamination resulting from wrongful use of chemicals was expressed by one of the officials of the Agricultural Extension Services. He mentioned that in efforts to control resistant diseases and pests, some urban cultivators concoct chemicals which might in the end be dangerous to humans. According to this official, the average urban farmer does not know much about agricultural chemicals so this is a dangerous practice.

Mosha (1991) notes that uncontrolled livestock-keeping practices within urban areas compromise city health. An example of some public opinion about keeping of livestock in towns and cities will suffice. Criticizing a neighbour for compromising city health by keeping dairy cattle in New Delhi, some inhabitants said,

> None of us like this man. Look at what he has done to our neighbourhood. We can't even open our windows, the smell is too bad (quoted in *Globe and Mail,* 2nd February 1994).

Apart from the fear of bad odour from animals, some officials fear that human diseases may be spread by livestock in urban areas. This is what an official of Ghana's Department of Parks and Gardens said,

> You see, animals in the city [Accra] are sources of bad odour. In addition, there is always the risk of spread of diseases by animals roaming the streets.

All the above-mentioned problems may be 'caused by poor practice through lack of information and extension assistance' (UNDP, 1996:199). Consequently, as an official of the Policy Planning Monitoring Evaluation Department (PPMED) of the Ministry of Agriculture put it, the problems may be alleviated if urban farmers are given proper education about the use of chemicals and about good animal husbandry.

It is argued in some official circles that agriculture should not be allowed in cities since 'production of food in the polluted environment of cities may cause contamination' (UNDP, 1996:199). This concern may be valid but banning urban agriculture because of industrial pollution does not solve the problem of pollution. I asked the farmers for their response to this concern. In sum they said, apart from the possibility of affecting crops, industrial pollution may directly affect human beings, the soil, water and air in cities. Consequently, instead of banning urban agriculture, industrial pollution should be checked.

In June 1987 the Rift Valley Provincial Commissioner of Nakuru in Kenya said, 'maize farming in the town of Nakuru would not be allowed because thugs hid in the shambas... The Administration would deal with the few who had decided to plant maize on plots in the town' (*East African Standard,* June 2, 1987).[1] In Tanzania, 'urban by-laws ... empower town and municipal authorities to destroy crops grown within the urban centres which are a metre high' (Mlozi, Lupanga and Mvena, 1992:287).[2] Also in Bamako (Mali), the

[1] Shamba means garden.

[2] Supposedly because of the notion that it harbours thugs or that it taints the beauty of Tanzanian towns and cities.

authorities have banned the cultivation of cereals since 1989 on the grounds that the tall stalks provide hiding places for bandits (Diallo, 1993). During this study, a man cultivating on Ghana Broadcasting Corporation's (GBC) land said,

> Officials of GBC say we should not cultivate tree crops because they create shadow [hideouts] for criminals.

However, the grounds for banning the cultivation of cereals in some cities in sub-Saharan Africa are not firm. During an interview with one of the officials of the Extension Services of the Ministry of Agriculture I asked whether he thought the grounds for banning urban agriculture in Mali, Kenya and other such countries were reasonable, and whether it was good grounds to ban urban cultivation in Ghana. He said,

> Banditry is in the mind of people so if bandits cannot hide in cereal crops they would hide elsewhere. This means the banishing of cereal cultivation would not prevent banditry. Apart from cultivated crops, bushes grow in African cities, consequently, these as well can provide hiding places for bandits.

It should be noted that due to the poor economic conditions of many sub-Saharan African countries, city authorities do not have enough resources to keep African cities free of bushes. This statement was confirmed by an official of Accra Metropolitan Assembly when she said that the Assembly is not able to clear Accra of all bushes so 'we should be grateful to urban cultivators for clearing the bushes.' It follows that it is better to encourage urban residents to clear and cultivate unused areas in the cities rather than allow the areas to grow into bushes or weeds.

Another reason for official skepticism to urban agriculture is the assertion that some cultivated crops, like stalks of maize, provide breeding places for mosquitoes. The argument is that water gathers in the axils of maize plants thus, creating breeding places for mosquitoes. This assertion was reiterated by an official of Parks and Gardens in Accra. It has, however, been proven by research conducted by Watts and Bransby-Williams (1978) in Zambia that mosquitoes do not breed in maize stalks.

According to Goodland, Watson and Ledec (1984), public health problems in the tropics, in which mosquito-related diseases alone afflict millions, make the use of biocides for disease vector control unavoidable. However, the widespread use of biocides in agriculture is known to result in the emergence of resistant strains of mosquitoes and other disease vectors. For example,

> By 1976, 43 species of anopheline mosquitoes (vectors of malaria) throughout the world had developed resistance to dieldrin, and 24 species were also resistant to DDT. Resistance to these biocides by culicine mosquitoes (vectors of yellow fever, encephalitis, filariasis, and dengue) increased from 19 species in 1968 to 41 species in 1975 (Goodland, Watson and Ledec, 1984:138).

Since resistant mosquitoes emerge as a result of the use of biocides, it can be asserted that the incidence of malaria has gone up in areas of heavy biocide use. And it is reasonable to conclude that urban agriculture is an important contributing factor to this trend. However, the best solution to this problem may not be banning urban agriculture. I asked all the officials involved in this study this question; 'you agree that urban agriculture is very vital for the people involved, and for the population in general. However, it has been confirmed in some circles that the increased use of biocide in agriculture results in resistant mosquitoes. This means an increase in malaria. Do you think we have to ban urban agriculture for this reason?' The answer provided by the official of the Waste Management Department of Accra Metropolitan Assembly sums up their answers. He said,

> Instead of banning urban agriculture, farmers should be educated in the use of other means, especially biological, to control pests and diseases.

In many sub-Saharan African cities, water supply is a problem. As a consequence, urban farmers compete with other users for available water. In this struggle, urban farmers tend to lose to the other water users because the fee for tap water is very high. Urban farmers, especially open-space farmers, therefore make use of other sources of water including untreated gutter water. As noted elsewhere, African cities have 'no local treatment facilities or standards and monitoring systems to ensure the purity of waste water before it is applied to land crops' (UNDP, 1996:216). Therefore urban cultivators are accused of using waste and polluted water to water their crops. The use of unwholesome water by some urban cultivators has prompted some officials to be skeptical about the practice. It should be noted that the use of gutter water poses a real danger to human lives. This concern was raised by an official of

Ghana's Parks and Gardens when he said uncontrolled use of waste water in urban agriculture poses a threat of cholera outbreak. A typical example is the outbreak of cholera in Santiago, Chile in the early 1990s as a result of the consumption of 'tainted vegetables, grown in metropolitan Santiago using irrigated water polluted by raw sewage' (Bartone, 1994:12). Ghanaian officials involved in this study suggested that to stop the use of gutter water, groups of urban farmers should be financially and materially assisted to sink boreholes in their areas of farming. This may entail the formation of cooperative organizations to provide services that individuals cannot afford on their own.

The more affluent farmers like enclosed farmers do not use waste water as they can afford tap water. It should, however, be mentioned that some officials do not support the use of tap water for farming. The following is what an official of Ghana's Extension Services had to say,

> Very few [cultivators] can afford to pay for water bills..., even the use of tap water is not advisable especially when the city is growing and our infrastructural layout is so poor. If we keep on tapping [water] at will, uncontrolling for something that harvested water can do then I don't think we are being fair to the economy.

Furthermore, some officials are hostile to urban agriculture because like other informal sector activities, 'it is thought not to abide by zoning and licensing laws' (House, 1978:38). Urban agriculture does not conform to zoning and licensing laws, since in planning sub-Saharan African cities the colonial administrators ignored urban agriculture, and 'there has been remarkable continuity from colonial practice in this sphere across the continent' (Simon, 1992:145). Furthermore, as urban planners and developers tend to associate development with industrialization, they tend to ignore agriculture in urban areas. The planners see food production as generally 'external to the cities, in peri-urban belts or adjacent rural food cachment areas' (Guyer, 1987:15). It therefore follows that generally in sub-Saharan Africa, urban agriculture does not have any impact on city planning, and vice versa.

Similarly, in many African countries 'little is formally known about urban agriculture' (Maxwell and Zziwa, 1992:3). As Freeman (1991) put it, the cultivation of food crops on a large scale in the public and private open spaces of cities, especially in the developing world, is common but remains almost untouched as a topic of research. With little research on the practice it is not surprising that some officials view it with discontent.[3] In line with this view is the fact that many urban residents are not conscious of the existence of urban farming. When I explained my research interest to some urban residents in Accra many asked, 'did I hear you correctly?' Incidentally, in the end these same people recommended some cultivators to me for interviewing. They know urban residents who cultivate in the city but they have never thought of them as farmers.

It is pertinent to mention that in Ghana, urban agriculture does not seem to fall within the jurisdiction of any state department or Institution. The reason is that it is considered part of the informal sector, agriculture which falls outside mainstream agriculture, and the government does not regulate the activities of this sector of the economy. Apart from the Agricultural Extension Services, all the institutions I interviewed had no plans or agenda for urban agriculture. And it is yet to be seen, whether the Extension Services' agenda for urban agriculture would be implemented. The question is, why have they not implemented any significant agenda in the previous years?

Since urban agriculture does not fall within the jurisdiction of any state department, it does not have any official representation and consequently no official organ to cater to it. In Kenya, Lee-Smith and Memon (1994) note that even though urban cultivators constitute a substantial portion of the urban population, they are yet to be represented by any organization, either in any town or at national level.

During my fieldwork in Ghana, I realized that many officials did not know much about urban agriculture. Consequently, they did little or nothing to improve upon urban agriculture. For example, in response to my request for an interview, one of the top officials of the Accra Metropolitan Assembly (AMA) asked me, 'what has urban agriculture got to do with us?'[4] When I questioned officials regarding the number of people involved in urban agriculture in Accra, the majority of them told me they did not know. In fact, some even questioned why I would expect them to know. The official from the Department of Parks and Gardens said, 'almost all those in Accra who have idle land, and also those who do not have adequate income are involved in urban cultivation.'

---

[3] It should be noted that it has now attracted a lot of researchers.

[4] AMA is responsible for Accra. Its duties include the provision of 'adequate and appropriate parks, gardens and avenues of trees in suitable areas,' and 'to ensure with other agencies proper city planning and zoning' (AMA official).

On the other hand, the officials of the Extension Services of the Ministry of Agriculture predicted the number of part-time urban farmers in Accra to be 10 to 15 per cent of the population and about 5 per cent involved on full-time basis. The officials of the Extension Services emphasized that their figures were not accurate, as they have not conducted any surveys to establish the correct number of urban farmers. While the officer from the Department of Parks and Gardens gave the impression that the majority of the urban residents in Accra may be involved in urban agriculture, those of the Extension Services gave the impression that only a minority were involved. This disparity highlights the fact that Ghanaian officials do not know much about the practice of urban agriculture.

It is true that some officials are in fact hostile to urban agriculture because they don't know much about it. Similarly, some officials discourage urban agriculture due to associated hazards. However, there is an important factor that previous researchers neglected as far as official attitude toward urban agriculture is concerned: this is an issue of the social class of the farmers. Social class and power positions of urban farmers play an important role in the acceptance or rejection of the practice. Earlier studies 'assumed that UA was done mainly by the poor, uneducated, and unemployed men and women in urban squatter areas' (Sawio, 1994:26). If urban agriculture was mostly practiced by women and the poor as some researchers have found in various countries, then one might infer that official resentment toward the practice was due, at least partly, to the low status of the farmers.

Since most previous researchers did not study the characteristics of urban farmers over a period, of time they could not report to us how gender and class position have affected official attitude toward the practice over the years. The few who did study the characteristics of urban farmers (over a period of time) did not connect this to official attitude toward urban agriculture. However, an important assertion made by one researcher is worth noting. Sawio (1993) has stated that the increased involvement of highly educated people in urban agriculture may help legitimize it. His reasoning was that,

> The more educated the players in the enterprise, the more likely will they be interested in protecting their investments by influencing policies and regulations in its favour (Sawio, 1993:19).

An analysis of my data indicates that in Accra before the 1970s most of the farmers were night watchmen (security-men), garden boys, unemployed, illiterates, and so on.[5] There were only a few people from the middle/upper class involved in the practice. Similarly, more women than men were involved in the practice.

A small explanation of why more women than men were involved in the practice in the 1950s and 1960s would be appropriate. Comparatively, in the late 1950s and 1960s there were a lot of unskilled work positions so many urban residents, especially men, were employed. When Ghana attained political independence from colonial rule it embarked on industrialization, thus setting up many industries which absorbed a considerable number of the inhabitants, especially men. Since a large number of men were employed, and since in many cases they had to work long hours, they did not have time to cultivate (see chapter 2).

In the 1950s and 1960s the government (through purchasing and distributing corporations) and large private establishments like United African Company (UAC) and GB Olivant (Ghana Ltd.) monopolized the distribution of consumer goods. As a result, many urban women, especially housewives, did not earn sufficient income from the distribution sector. Secondly, by design, African women who were officially employed were employed in shorter shifts, and in low paying positions. This is indicated in the recommendations of a Commission set up in the late colonial period. The Commission suggested that,

> Apart from posts such as teaching in girls' schools, midwifery, etc., it is, generally speaking more economical to employ women than men on jobs which involve work of a routine or manipulative and repetitive character not involving long and expensive training, and which offer only limited prospects of advancement. We therefore recommend that the Government should take such steps as are practicable to attract educated women into the Civil Service at all levels, but practically in posts such as typists, stenographers, machine operators, and clerical assistants. We consider that, other things being equal, preference should be given to women candidates for such posts (Commission on the Civil Service in Gold Coast, 1951:44).

Generally, women did not have variant sources of income. In addition, they were concentrated in low paying positions and did not work much over-time.

[5] Officials, cultivators and non-cultivators all confirmed this assertion.

Consequently, many women engaged in agriculture mostly to supplement their food supply. Taking into consideration that at that time jobs dominated by women tended to be of low status, it is not surprising that urban agriculture was looked down upon and discouraged. When women were the majority in urban farming the practice was seen as 'a form of recreational pastime for housewives or a minor and unimportant domestic activity like doing the family washing' (Freeman, 1991:79).

Since 1972 many people of higher social class have been involved in urban agriculture.[6] I asked cultivators of the first time they got involved in urban agriculture. For their answers see the table below.

Table 5.1 First Time the Cultivators Got Involved in Urban Cultivation

First Time in Cultivation	Frequency	%
before 1966	20	10.0
1966-75	67	33.5
1976-85	55	27.5
1986-95	58	29.0
Total	200	100.0

Figures in Table 5.1 indicate that most of the farmers in this study got involved in the practice for the first time between 1966 and 1975. An important item of information, not clear from the above table, is that most of the people who got involved in urban agriculture within this period began in 1972, when Ghana embarked upon Operation Feed Yourself (see Chapter 4). A more important observation from Table 5.1 is that most of the farmers got involved after Ghana attained political independence in 1957, and after the first Ghanaian government was changed in 1966. My impression is that the first Ghanaian government after independence was more interested in maintaining the beauty of Ghanaian towns and cities as determined by the colonist. Moreover, comparatively, the economic situation in Ghana was better from the

[6] Since there are no data on this I will rely on people's perception of changes in the social class of urban farmers.

time of independence to the time the first indigenous government was toppled. A combination of good economy and stringent government prohibitions discouraged many urban residents including the middle/upper class from engaging in agriculture.

Now, I will examine why officials are increasingly becoming more positive to urban agriculture, starting with socio-political reasons.

5.3 Reasons for Official Positive Attitudes Toward Urban Agriculture

There are two main reasons why officials have positive attitudes toward urban agriculture. These are socio-political and economic factors. I will begin with the socio-political factors.

In Ghana, the increasing involvement of higher class citizens in urban agriculture helped to induce government officials to take on a positive attitude toward the practice. Politicians, professionals, business people and managers are very influential so government bureaucrats would not want to antagonize them. Their involvement in urban agriculture has raised the prestige of the practice and has affected the way government officials view the practice. This has been observed in Tanzania also. In Dar es Salaam, for example, it has been noted that,

> Attitudes towards livestock in the city changed considerably when the richer residents of Oyster Bay started to raise cows in their backyards (UNDP, 1996:212).

Also, by virtue of their educational background, their position in the decision making apparatus, and their access to economic resources, many top government officials are often classified as middle/upper class citizens. Many government officials eventually got involved in urban agriculture and it is rational to assume that people do not make decisions that affect them adversely. Thus middle/upper class citizens/government officials including politicians would not make decisions or take actions that are not in their own interest.

Consequently, as many government officials got involved in urban agriculture they tended to view the practice in a positive way. About 75 per cent of the officials I interviewed said they are involved in urban agriculture in one way or another. In addition, they stated that if they did not see agriculture positively they would not have got involved in it. The others who were not involved in urban agriculture during the time of this work said their lack of involvement was due to time constraint or lack of land around their houses for agriculture. They are not against the practice, actually, they have friends and colleagues who are involved in urban agriculture. Some middle/upper class urban residents in Accra are so enthusiastic about urban agriculture that they claim the practice beautifies the landscape, prevents surroundings from going bushy, and helps drive away snakes and other poisonous creatures.

The increase in the use of universal franchise to elect political leaders has also contributed to the condoning of urban agriculture by Ghanaian politicians. The 'one-man one-vote' phenomenon has empowered average Ghanaians, especially urban residents. In order to win their votes, politicians are increasingly accepting illegal but harmless activities like urban agriculture performed by urban residents.

Public opinion has also played an important role in official acceptance of urban agriculture. Over the years, the general public has changed its attitude toward the practice. During my fieldwork I solicited the views of nonfarmers about urban agriculture. On this issue they were asked, 'do you think urban agriculture should be encouraged?' All forty respondents answered in the affirmative. The nonfarmers were further asked, 'would you have given the same answer some 30 years ago?' Over half of the respondents said 30 years ago they would not have suggested that urban agriculture be encouraged. The remaining half did not know what their answers would have been 30 years ago. Most of those who said they did not support agriculture in urban areas 30 years ago did not have any concrete reason. The words of an elderly woman sums up what they said,

> Thirty years ago? Let me see... which year are we talking about? Oh, 1966. At that time, I just felt agriculture in the city was bad. I cannot really tell you why I felt that way but it may be because of my experiences when I was a young girl. I grew up at Koforidua, and my mother had a nice vegetable garden in front of our house right at the center of the town. One morning, Town Council officials came to slash down every crop saying it was forbidden to cultivate crops in the town. In our neighbour's front-yard was growing flowers. These were not slashed down. Thinking of it today, we were allowed to grow flowers but not vegetables. I grew up believing cultivation of crops in urban areas was bad. Today, I think otherwise. What use is it growing flowers instead of vegetables? We need food not flowers.

The above observation suggests that, due to colonial influence, prior to 1966 many urban residents did not support urban agriculture, however, after 1966 they did. Further questioning revealed that most nonfarmers had a positive attitude toward urban agriculture for the first time in 1972. This was the time Operation Feed Yourself was launched. This was the time urban agriculture became an issue in Ghana (see Chapter 4). The nonfarmers gave various reasons for their attitude change toward urban agriculture. Among these reasons were the following:

1. Urban agriculture beautifies the cities—the municipal assemblies are not able to clear the weeds growing in the cities but urban cultivators cultivate such areas making them appear neat;
2. When people are hungry is it not reasonable for them to cultivate empty spaces for food? This is not our tradition—we don't go hungry when there is land to till; and
3. We need vegetables in our daily diet and since we cannot afford refrigerators, urban residents have to cultivate these themselves so that vegetables can be readily available in their fresh state.

In line with the above observations is an increase in the number of affluent people in the Ghanaian society, and subsequently an increase in the demand for vegetables and other nutritious foodstuffs. It is known that, 'food consumption is greatly influenced by income levels and food prices. On average, consumption of calories increases with income ...' (Braun et al, 1993:18). This observation is supported by Grigg (1995). He notes that when a person's income increases or when they become more affluent, they decrease their intake of starchy food and buy more expensive food like fresh vegetables and fruits (Grigg, 1995). It follows that as people's income increases they tend to get more of their calories from livestock products and from more expensive crops.

The increase in the number of affluent people in Ghana may be due to the open economy policy, including the privatization drive of the current government. In a typical third world country, it is not the number of the affluent that counts but rather their power. Many government officials and other decision makers are getting richer and subsequently demanding more vegetables. Since they get most of their supply of vegetables from within urban areas, they approve or condone urban agriculture.

That implicit public approval of urban agriculture is one of the factors that influenced change of official attitude toward the practice was confirmed by eight of the officials I interviewed. They asserted that if the general public was supportive of the practice then they did not have to discourage it. Signs of public approval include urban residents' increased purchase of urban grown crops, especially at the places of cultivation, and an increase in the number of urban farmers. An official of Accra Metropolitan Assembly mentioned during an interview that many people go to the farms to purchase crops because they don't see anything wrong with urban cultivation. He went on to say that officially nobody has ever complained about urban agriculture as a nuisance.

Government intervention, in this case indirect, is another factor that encouraged official recognition of urban agriculture in Ghana. A good example is the launching of Operation Feed Yourself by the then military government in 1972. This is discussed in chapter 4.

In 1992, the personal intervention of the President of Ghana helped to further encourage positive attitude of officials towards urban agriculture. In that year, officials of the Department of Parks and Gardens gave 'stop cultivating' orders to a group of cultivators at a place near the Osu Castle in Accra, the seat of government. This is what one of the cultivators told me during an interview:

> When we received the order we came together and sent a petition to the President. We told him that we are law abiding citizens with no source of income aside from the income we get from farming. Since we are not rich enough to buy land to cultivate we cultivate public land near the Castle. Before we started cultivating the area it was bushy and many people used the place as their toilet. Now the Department of Parks and Gardens say it is their land so we should quit. What shall we live on if we stopped farming?
> We told the President that our initiatives should be appreciated by the Department of Parks and Gardens because some people like us were roaming the streets stealing and doing other illegal things, and we were living a decent life.

On receiving the petition, the President assembled the head of the Department of Parks and Gardens and the representatives of the cultivators. An agreement was reached, whereby the cultivators were permitted to farm a part of the area and the Department of Parks and Gardens agreed to maintain the other side. This division was made easier by the fact that the area of contention is divided by a big open ditch/gutter. The cultivators were permitted to farm the area behind the ditch. Since news of the President's intervention became public, officials do not ask cultivators to stop cultivating until an area is ready for development.

As earlier mentioned, some economic factors were instrumental in changing the negative attitudes of government officials toward urban agriculture. Some Ghanaian officials have always held the view that Ghana would eventually become more industrialized, and subsequently many workers would be needed in the industrial sector. In this regard, foreign investors would be more willing to invest in the country if they were sure of recruiting the needed labour without much difficulties. Therefore, to have prospective industrial workers (surplus labour) readily available it was necessary for some potential workers to live in Ghanaian cities despite the current and increasing rate of unemployment. Government officials allowed potential workers to sustain themselves through agriculture and other informal sector activities. As an official stated,

> If urban residents are banned from farming in the cities many would not be able to survive..., and may even abandon the cities. However, we shall need them when we eventually become more industrialized.

Allowing urban residents to farm in cities removes from the government and the capitalist the burden of maintaining the unemployed potential labour force. The official's statement above notes that if urban agriculture were not condoned, many potential workers may abandon cities. This seems more of a hypothetical statement so I will treat it as such.[7] If such a situation emerges, capitalists logically have one basic choice. They have to recruit their labour from rural areas. This has two options, both of which are expensive for the capitalist. The capitalists or their agents have to relocate, albeit temporarily, to in rural areas to recruit. The other aspect is for the capitalist to advertise for hiring in the media. However, it should be noted that the media is not well established in Ghana and many other developing countries, and newspapers do not reach many rural areas. Consequently, the second aspect is not a very effective choice.

A problem with hiring directly from rural areas is that, 'rural labour normally do not have the necessary skills to work in urban industries and other institutions' (an official of Policy Planning Monitoring Evaluation Department). This implies that to make skilled labour available, employers

---

[7] The urban unemployed will not leave the city for the country-side. Even an enticing package by the Rawlings' government did not convince retired and laid-off civil servants to move to rural Ghana.

have to spend a lot of money to train their workers.

For the government and other employers there is another advantage in allowing workers to produce some of their foodstuffs. It enhances the stability of the economy of Third World countries. Urban workers agitate for more salary when their exchange entitlement is low, and when they cannot easily afford basic commodities especially food. Consequently, by producing some of their food, workers may become less militant in their agitation for increased salary. Therefore, in the tradition of Marxist theory, urban workers in developing countries have false consciousness. Since they produce some of their food, they do not feel the realities of exploitation.

The implication from the above is that workers are more able to survive on a meagre salary when they subsidize their food purchases through farming. The capitalist, therefore, does not pay for the reproduction of labour in developing countries. This is possible because of 'the continued existence of non-capitalist structures which provide support for the laborer, but which are not maintained by the wage he is paid' (Curtis, 1995:21). In the tenets of conflict theory, the intention of controlling worker unrest is one of the reasons why government officials condone agriculture in urban Ghana.

The inclusion of vegetables, especially salad, in the diet of Ghanaians also compelled Ghanaian officials to acknowledge the existence of urban agriculture. But for a few greens like 'kontomire,' vegetables were not usual part of the average Ghanaian's diet. Consequently, the average person did not care much about vegetable cultivation. However, with increased general education and knowledge in nutrition in particular, many Ghanaians have become more conscious of the importance of vegetables in their diet.[8] The officials interviewed for this work were asked whether the countryside is able to supply enough food and vegetables to feed the urban population. They all confirmed that this is not the case. As an official of Accra Metropolitan Assembly put it,

> Owing to logistical constraints including the use of crude agricultural tools, poor transportation network, and lack of adequate refrigeration the countryside is not able to supply enough vegetables to the urban areas.

---

[8] Earlier, salad was regarded as vegetable for the affluent. Since it was not widely consumed it wasn't produced in any significant quantity in the country. It was therefore imported, thus making it very expensive.

From the above quotation, one may correctly conclude that in Ghana inefficient transportation system and other such hindrances have made it unrealistic for cultivators to do any meaningful vegetable farming in the areas around urban settlements, since they are not able to transport their harvested crops into the towns/cities in good time. Vegetables should be eaten in their fresh state, but for urban residents to consume fresh vegetables it is necessary for them to engage in urban agriculture. This is what an official PPMED said,

> One might say that agriculture should be limited to the country-side. But we should realize that we have to use vegetables in their fresh state. Therefore, the issue is that vegetables should be sold or bought when fresh. That means, there should be an efficient transportation system. That is not guaranteed so the only alternative left to us is to grow vegetables close to the market. Vegetable cultivation should be close to the market because most Ghanaians don't have refrigerators. They should buy vegetables on daily basis.

It should, however, be mentioned that farmers who cultivate export-oriented crops like pineapples around Accra have been able to build their own transportation systems. Incidentally, poor urban farmers are not able to afford such amenities. However, with an increased number of prominent and influential people moving to the urban periphery, and with an increased involvement of prominent Ghanaians in urban agriculture the transportation problem will eventually be solved.

An increasing foreign population in Ghana also necessitated the production of vegetables in urban areas where expatriates are concentrated. According to an official of the Extension Services of the Ministry of Agriculture,

> In order to retain and attract foreign investors we should be able to increase the supply of fresh vegetables in our cities. You know how whites depend on vegetables. Consequently, it is pertinent to allow agriculture in urban areas.

Prior to the launching of Operation Feed Yourself, Ghanaian officials wrote urban agriculture off as ephemeral. However, from their answers to some of my questions I gather that officials now realize the importance and permanency of this practice. In the field, the Ghanaian officials involved in this study were asked whether urban agriculture is a permanent or temporary practice. All of them believed that it is a permanent practice. The words of an official of the Extension Services of the Ministry of Agriculture sums up what the officials said about the permanency of urban agriculture. He said,

> Increasingly, many people are getting involved in urban agriculture. And those already in it are not abandoning it so you can say it is a permanent practice.

If urban agriculture has become a permanent issue then government officials should not try to discourage, but rather encourage it.

According to the officials in this study, the general food supply from the countryside has become inadequate. Conversely, the demand for foodstuffs in the cities has increased. Apart from one official, all the other officials noted that the inadequate supply of food from rural Ghana is a good reason for the government to encourage urban agriculture.

Similarly, there is a shortage of foreign exchange to import food, so it has become more necessary for urban residents to produce some of it. When asked about the importance of urban agriculture, all the 12 officials I interviewed mentioned, among others, that it saves foreign exchange meaning urban residents produce vegetables that the government would have spent money importing. Due to this and other reasons most of them felt it was necessary to encourage urban agriculture.

Though condoning of urban agriculture has increased over the years, not all animals and crops are tolerated in urban Ghana. When asked to list animals not appropriate to be raised in urban areas the officials mentioned pigs and cattle. They noted that cattle and pigs may be reared in peri-urban areas but not at core areas. Their reason as expressed by an official of the Extension Services is that,

> ... our cattle are reared on free-range so it's better to keep them away from places of high human density. Cattle can hurt people.

On the basis of this response, I asked the officials whether they would approve of cattle rearing in core areas of cities if they were confined. The general answer was no. In the words of one official,

> It is not appropriate to keep cattle in the midst of people because their pens emit bad odour, and there is always the danger of some of them escaping from their pens and hurting people.

Pigs are also not recommended for rearing at the core areas of cities because, as an official of Accra Metropolitan Assembly put it, they are dirty animals and they make too much noise.

The crops officials did not recommend for urban cultivation are traditional crops. They mentioned crops like plantain, oil palm and sugar-cane. Basically, the crops they did not approve for cultivation were crops that grow tall, took long to mature, and are easily cultivated in rural areas.

The crops officials declared appropriate for cultivation in urban areas are the same as those approved in some by-laws for cultivation in Ghanaian cities. The existence of these by-laws is an indication that there are some regulations or policies that regulate urban agriculture.[9] It is necessary to mention that these regulations date back to 1972 but they have been overlooked and are still being ignored by both officials and urban residents in general.[10] The existence of these by-laws is an indication that urban agriculture has been officially sanctioned in Ghana since 1972.

The regulations or by-laws as they are termed in the LOCAL GOVERNMENT BULLETIN OF 1972 (updated in 1976, 1977 and 1994) states in Act 359 that 'the maximum number of poultry that may be kept in a dwelling-house within the Council [Accra City Council] area shall be two hundred heads.' Act 359 also regulates the keeping of animals in urban areas under the captions 'Impounding of Excess Poultry,' 'Housing of Poultry,' 'Size of Housing,' 'Disposal of Droppings,' 'Stray Poultry Disallowed,' 'Power of Health Officer to Conduct Inspection,' 'Discretionary Powers of Accra Metropolitan Assembly,' 'Penalty,' 'Interpretation,' and 'Revocation.'

There are also by-laws on the growing and sale of crops. Since crop cultivation is of more importance to this work, I will focus on this aspect of the by-laws. The first by-law, also under Act 359, is captioned Requirements and Registration. It states that,

> No person shall grow crops at a place other than on land within premises unless he has registered with the Medical Officer of Health his name and address and the description of the site where the crops are to be grown.

This by-law, like the others mentioned below, is ignored by Ghanaian

[9] The regulations are not for the purpose of increasing urban food production. It is to maintain good sanitary conditions in the cities.

[10] 1972 was the year Operation Feed Yourself was launched.

officials. They admitted that they don't have any register over the people who cultivate outside their premises. It follows that urban cultivators do not register with the Medical Officer of Health, nor with any other authority. The second, Watering and Irrigation, states that,

> No crops shall be watered or irrigated by the effluent from a drain from any premises, or any surface water drain receiving water from any street.

As mentioned at the beginning of this chapter, some officials showed concern about the use of untreated gutter water by urban cultivators. However, despite the fact that many urban cultivators use gutter water, no attempts have been made to stop them from using it.

The third by-law is on Infected Persons. It reads,

> No person who has a discharging wound or sore or the symptoms of any infectious disease shall take part in the growing or sale of crops.

It is obvious that officials have no means of enforcing this by-law as they have little or no contacts with cultivators. Consequently, there is no way of identifying urban cultivators who have discharging wounds and/or infectious diseases.

Section four of the By-law is on Sale of Crops. It has four sub-sections as follows,

> 1. No crops shall be sold, offered or displayed for sale at a place other than in a market, stall, store or kiosk.
> 2. No crops shall be displayed for sale on a road, pavement or sidewalk.
> 3. The Medical Officer of Health may, where he considers it necessary in the interest of public health, declare any crops unfit for human consumption.
> 4. No crops declared unfit for human consumption shall be sold, offered or displayed for sale as human food.

My fieldwork revealed that, contrary to by-law Act 359 Section 4, the cultivators sell most of their crops on their farms rather than in the market, stall, store or kiosk. Officials are aware of this practice but have not attempted to stop it. On sub-section three, one may ask: how can the Medical Officer of

Health be able to identify crops unfit for human consumption when growing and/or harvested crops are never inspected?

The fifth by-law, Offences, states that,

> Any person who contravenes any of these by-laws shall be guilty of an offence and liable on summary conviction to a fine not exceeding C50.00 or in default of the payment of the fine, to a term of imprisonment not exceeding three months.[11]

Despite the fact that urban cultivators contravene the by-laws, no cultivator has ever been prosecuted. The most important reason is that no government institution enforces the laws.

The sixth and final by-law is on Interpretation. It states,

> In these by-laws, unless the context otherwise requires, crops means lettuce, tomatoes, radishes, onions, cucumber, water melon, oranges, bananas, *kontomire* [cocoyam leaves] or any other agricultural produce likely to be eaten in an uncooked state.

An implication from the sixth by-law is that mainly only vegetables and fruits are permitted to be grown in Ghanaian towns/cities. It may therefore be concluded that urban agriculture is officially allowed in Ghana because of, among other factors, the increasing demand for vegetables.

The foregoing discussion indicates that officially much is not done to improve and control urban agriculture—it is allowed to take its 'natural course.' It is therefore not taken into consideration when planning Ghanaian towns/cities. In addition, officials do not provide enough assistance, if any, to urban farmers. An official tried to explain away the lack of government involvement in urban agriculture saying,

> Ghana is a capitalist country, and the abstract model of capitalism demands limited direct government intervention in the economic system. As a result, we are not ready to make any major interference in urban agriculture. We should limit our intervention to protecting farmers and the general public

[11] C50.00 is less than US$0.50 in 1998 currency. It is possible the fine has been modified but there is no document to support this assertion.

> from contamination and any other ills associated with urban agriculture.

This seems to be the idea of many officials. One of them mentioned that the informal sector functions best without government intervention. Therefore, Ghanaian officials advocate a limited government intervention in urban agriculture.

For a specific period of time, the official attitude towards urban agriculture may be influenced by major events like international conferences. Consequently, I asked the officials involved in this study the following question; 'what is the normal government attitude toward urban agriculture during important events like the recent PANAFEST or Non-aligned meeting in this city?' The answer given by an official of the Extension Services of the Ministry of Agriculture sums up their response,

> In a general sense, official attitude toward urban agriculture does not change. If some crops are in the way of a conference they will be destroyed. Otherwise, everything goes on normally.

This answer is similar to what was found out in Kenya. According to Freeman (1991), crackdowns on unsightly jua kali enterprises and urban shambas intensify when major international conferences in the city bring in large numbers of foreign visitors. In a reaction to the possibility of destroying cultivated crops in Ghanaian towns and cities where international conferences may be taking place, I asked the official whether cultivators would receive any compensation if their crops were destroyed by the government. The answer was no. In Ghana, when a property like a building is expropriated for a government project, the owners receive financial compensation or restitution from the government. Since this is not extended to crops cultivated in cities it means that urban agriculture is not protected by law.

I wish to make a general observation on the types of crops sanctioned by Ghanaian officials on the field and in by-laws. There is not a radical shift from colonial regulations. The crops officially allowed to be cultivated in urban Ghana are those mostly consumed by middle/upper class Ghanaians and expatriates. Carbohydrate crops like cassava and plantain consumed mostly by the lower class are prohibited. It is the same issue of those in power satisfying their needs at the expense of others. The colonial administration banned urban agriculture in order to satisfy their needs, i.e. to beautify cities, and to discourage mass migration of Africans into the cities. Past Ghanaian authorities banned the cultivation of certain crops and the rearing of certain animals to satisfy their needs, i.e. to beautify cities. Present Ghanaian officials encourage the cultivation of certain crops to satisfy their taste, not that of the general population, and they discourage the cultivation of crops mostly consumed by the lower class.

5.4 Summary

In this chapter, I showed that government officials have changed their attitude toward urban agriculture from negative to positive. Prior to that, I discussed the reasons why they had negative attitudes toward urban agriculture. The factors discussed include the fear of spread of diseases by livestock, bad odour from livestock, hazardous effects of agricultural chemicals on humans, contamination of soil, crops and water by the chemicals used, and the emergence of resistant mosquitoes as a result of the use of biocides. The other factors discussed are the belief that urban agriculture provides breeding places for mosquitoes, that crops like maize in towns and cities provide hiding places for bandits and thugs, the fear that since some urban cultivators use untreated gutter water they expose urban residents to diseases, and that urban areas are polluted, so crops grown there are also polluted and consequently unfit for consumption. I also noted that officials did not condone urban agriculture because until the early 1970s most of the people involved in the practice were of low social status, most officials did not know much about the practice, and some officials strongly believed that agriculture was best practiced in rural areas.

I also discussed the factors that changed official attitudes toward the practice. Such factors include difficult economy/high cost of living, the approval of urban agriculture by most urban residents, the involvement of higher status individuals, and an increase in the demand for vegetables. Other factors that influenced officials to change their attitudes toward urban agriculture are government intervention, increase in the number of expatriates, the necessity to keep surplus labour in urban areas in anticipation of hiring them, the increased use of universal franchise to elect political leaders, and the need to maintain a stabilized economy.

To understand the people involved in urban cultivation, I discuss the characteristics of the cultivators in this study in the next chapter.

6 Characteristics of Urban Farmers

6.1 Introduction

There is a consensus among various researchers on urban agriculture that various types of people are involved in the practice—both young and old, males and females, rich and poor, literates and illiterates. Some urban farmers are employed in the formal sector and others are not (Lee-Smith and Memon, 1994; Mougeot, 1994; Sawio, 1994; Maxwell, 1993; Maxwell and Zziwa, 1992; Freeman, 1991).

Urban farmers give different reasons for farming. Depending on their characteristics, the reasons can be summed up as economic and socio-cultural. As mentioned in chapter 2, it is mostly economic reasons that compel urban residents to find jobs in the informal sector. However, choosing urban agriculture instead of other informal sector jobs is determined by cultural factors.

Here, I will examine the general characteristics of urban farmers in Accra, and in the next chapter I will examine differences between the two main types of farmers identified in this study. For now I propose that urban residents who have lived in Accra for a considerable length of time are more likely to be involved in urban agriculture. This is based on the assumption that the longer a person stays in an area, the more familiar he/she becomes with the land tenure system; he/she becomes aware of land lying idle, who to contact for a plot of land, and the like.

Another proposal that will guide me in examining the characteristics of urban farmers in Accra is: the larger the family, the more likely they are to farm in the city. Holding other factors constant, it can be said that the larger the household, the greater the need for food and consequently, the more pressure to farm.

In chapter 2, I emphasized the importance of the rural background of urban farmers. Thus, I expect most urban farmers to have rural origins and/or lived in rural areas. Therefore, I propose that most urban farmers have previous knowledge of agriculture. On the issue of age, I expect most farmers to be of middle-age. All things being equal, younger urban residents spend time in the hope of securing jobs in the formal sector, so do not engage in urban agriculture. Thus, there should be fewer urban farmers who are very old/young. The very old may be too fragile to engage in farming. Also, I expect more men than women to be involved in urban agriculture. This expectation is based on the fact that Ghanaian, indeed West African, urban women prefer trading to urban agriculture.

Lastly, I propose that urban residents who get involved in urban agriculture see the practice as a permanent issue (see chapter 2). Until compelled by forces beyond their control, urban farmers will not quit farming.

6.2 Gender, Age and Marital Status

It is impossible to generalize whether men or women form the majority of urban farmers in sub-Saharan Africa because it differs from region to region or from country to country. Some researchers note that the majority of urban farmers in Africa south of the Sahara are women (see Tinker, 1994; Maxwell and Zziwa, 1992; Freeman, 1991). Confirming this assertion, studies in some towns in Kenya showed that up to 63 per cent of the cultivators were women (Lee-Smith et al., 1987; Freemen, 1991). Further, studies of other East African countries have also shown that women form the majority of urban cultivators (Sawio, 1994; Mvena et al., 1991; Tripp, 1990; Rakodi, 1988). This reflects 'women's traditional roles in the production and reproduction [of] the family' (Sawio, 1994:30). In many places in sub-Saharan Africa there is the 'perception that women ought to cultivate because it is the wives' duty to provide the household with food' (Sanyal, 1984:11). These observations in East African cities affirm the most important tenet of the labour-surplus model mentioned in chapter 2, that surplus labour has rendered the less powerful people (especially women), jobless thus compelling them to engage in urban agriculture.

Despite the findings mentioned in the preceding paragraph, it is not accurate to say that women form the majority of urban farmers in Africa south of the Sahara. Based on available evidence from past research, it seems more appropriate to say that the majority of urban farmers in East African cities are women. Even in East Africa, this generalization would still not be accurate. This argument is supported by Freeman's (1991) assertion that, 'it would be inaccurate... to label urban agriculture as a sector of exclusively female employment in view of the fact that male cultivators are a significant minority and in some areas they either equal or outnumber women' (Freeman, 1991:56).[1]

Contrary to the general notion regarding the gender of urban farmers in East Africa, in West Africa most urban farmers are men. The present work found that most of the urban farmers studied in Accra are men. See Table 6.1.

Table 6.1 Gender of Farmers in the Accra Study

Gender	Frequency	%
Male	139	69.5
Female	61	30.5
Total	200	100.0

The above table shows that 69.5 per cent of the farmers in this study are men, and only 30.5 per cent are women. The perception of urban farmers in Accra affirms the position that men form the majority of urban farmers. When I asked the farmers whether they thought the majority of the farmers were men or women, almost all said men form the majority. Thus, even the female respondents agreed that the majority of urban farmers are men. When I asked non-farmers the same question I received the same answer. The following is what a woman non-farmer told me,

> Can't you see it yourself? Wherever you go you see these men cultivating. You hardly see women among them. I will not cultivate in the city. Well, I do cultivate behind my house. What I mean is that I'll not cultivate in the open. I don't have to cultivate for sale, you know. I'll rather engage in retail trading than farming for sale.

Explaining men's omnipotence in urban farming, the female cultivators in

[1] Freeman did his research in the East African country of Kenya.

this study said that women are not able to compete favourably with men for land in the cities. Generally, 'urban women face greater difficulty than men in attaining access to land, water, credit, extension services and essential inputs' (UNDP, 1996:68). In addition, as earlier mentioned, Ghanaian women prefer retailing to agriculture. A woman cultivator in this study commented, 'it is only we, the poor ladies who engage in urban agriculture. The rich ones are businesswomen retailing one commodity or another.' An assertion by Maxwell (1997) confirms this by example of Ghanaian women in Accra. He notes,

> Urban retail marketing and petty trading are sectors that have long been dominated by women in West Africa, so it is not surprising to note that these are the most common forms of female livelihood in Accra (Maxwell, 1997:17).

In Senegal, another West African country, most urban farmers are men (Ratta, 1993). On one of the consequences of urban clustering in Ile-Ife, it has been observed that 'such urban clustering facilitated women's trading facilities and may help explain their relatively low participation in agriculture ...' (Tinker, 1997:126). These observations refute, at least in West Africa, the aspect of the labour-surplus model which indicates that due to their poor economic positions women form the majority of urban farmers. However, it does not refute the main tenet of the model, which stipulates that many unemployed are involved in urban agriculture.

During the colonial period and to some extent the early part of political independence, women formed the majority of urban farmers in West Africa due to, among other factors, their limited educational background and the fact that they were discouraged from dwelling in towns and cities. This prevented them from securing formal sector jobs. As noted elsewhere, women in the Civil Service were restricted by their lower level in education (Commission on the Civil Service, 1951:44).

Furthermore, in Ghana at a certain stage during the colonial period,

> ... the African men were opposed to employment of women in the Civil Service. This was based partly on the fear that, women, with fewer financial commitments, will accept lower salaries than men, who will, as a result be unable to find work (Commission on the Civil Service, 1951:44).

Table 6.2 Percentage Female Enrollment in University and Equivalent Institutions*

Country	1970	1980	1985	1988	1989	1990
Botswana	27	35	45	-	40	44
Central African Republic	3	9	11	15	11	15
Ghana	13	20	17	18	18	22
Kenya	-	25	29	-	31	28
Madagascar	32	-	38	43	44	45
Nigeria	14	-	24	27	27	-
Sudan	-	27	39	-	41	48

Source: Unesco's 1995 and 1996 Statistical Yearbooks.
* means no figures available.

In addition, most urban women were housewives and able to devote some of their time cultivating while their husbands worked outside the home.

Apart from their skills in distribution, women's education in West Africa (indeed, in sub-Saharan Africa) has vastly improved (see Table 6.2).

Due to the increase in their level of education, sub-Saharan African women are no longer only housewives, as was the case in the past. Their improved educational standard has increased their representation in well-paid formal sector employment.

As noted above, currently in West Africa most of the farmers are male. The primary reason, as mentioned, is that in this particular region the women are well skilled in petty retailing, therefore, engaging in distribution or retailing rather than in urban agriculture, especially since they find it difficult obtaining farmland. It is noted by an observer, 'indigenous marketing was the activity of... women' (Guyer, 1987:31). For example, in Accra as early as 1969, 83 per cent of traders in Accra were female (Reusse and Lawson, 1969).

The current redundancy in the formal employment sector has also contributed to men forming the majority of urban farmers. In the first place, due to higher gender inequality, i.e. higher discrimination against women during colonialism and early independence, more men than women were employed in the public sector. It was therefore expected that during the current restructuring more men than women would be laid off. Many men laid off from formal sector jobs turn to urban agriculture because they encounter more difficulty in securing readily available informal sector jobs, apart from farming. Conversely, it is easier for women laid off from formal employment to find women's work. Secondly, due to their rural backgrounds and the Ghanaian mentality that Agricultural Science is for men, many males do have farming skills.[2] Consequently, they resort to farming when they are laid off from formal sector employment.

Approximately 65 per cent of the urban farmers I interviewed were married. Twenty per cent had never married, and 15 per cent of my respondents were widowed or divorced. In five cases the divorce or death of a spouse led to lower income, consequently forcing them into agriculture. As one man put it,

---

[2] In Ghanaian places of formal learning like Secondary Schools, males are encouraged to learn Agricultural Science, and females to learn Home Science. Conscious that women produce most food in rural sub-Saharan Africa, I do not seek to imply that men are more skilled in farming than women.

> When my wife was alive I never dreamed of cultivating in addition to my tight-scheduled work because she made a lot of money from her retailing to assist the family financially. Now she is gone and I have to feed all our four children all alone.

One aged farmer cultivates because his wife is terminally ill. He said,

> Look at how old I am. I don't have to till the land to survive because I'm fragile. I did not do this when my wife was healthy. She has been very sick for sometime now. And we have spent all our money on hospital bills and on medicines but she is still not well. When she was healthy she did some petty trading that brought in enough money but now... well what can I do?

The majority of the respondents live in a revised type of nuclear family system. This consists of a couple and their children, and one or more relatives. This could include close and/or distant relatives such as nieces, nephews, and parents (see chapter 8).

Most of the cultivators in Accra (39 per cent) are between the ages of 35 and 44 years. This is followed by cultivators in the 45-54 age bracket (29 per cent).

There are fewer urban farmers above 65 years, and between 15 and 24 years of age. The fact that fewer farmers are in the younger age category confirms the assertion that new migrants to urban areas spend some time looking for wage employment at first. Most of the young will still be looking for wage jobs, while others might have given up the search and entered into urban agriculture. Alternatively, most of the young are still in school.

Figures in Table 6.3 show that over 50 per cent of the farmers in Accra are in the 35-54 age group, particularly in the 35-44 age group. This does not differ very much from findings among urban farmers in other African countries. In Nairobi, most urban farmers appear to be in their thirties or older—the median age is thirty-eight years (Freeman, 1991:56). Also in Kampala, Maxwell and Zziwa (1992) noted that the average age of respondents was 42 years. In Dar es Salaam, Tanzania Sawio (1994) found that over 50 per cent of urban cultivators of both sexes are in the 26-45 age group, only a few farmers in the higher age groups.

Table 6.3 Age Distribution of Urban Cultivators

Age Group	Frequency	%	Cumulative %
15-24 yrs	8	4.0	4.0
25-34 yrs	30	15.0	19.0
35-44 yrs	78	39.0	58.0
45-54 yrs	58	29.0	87.0
55-64 yrs	19	9.5	96.5
over 65 yrs	7	3.5	100.0
Total	200	100.0	

6.3 Size of Household

In this study, the concept of household means 'collective units that share housing and food, that trade clothing and other consumer durables, and that are composed of individuals who pool monetary resources' (Gottdiener, 1994:258). Not all members of a household need be immediate family. One's household may include their friends. Generally, in Ghana households are 'predominantly male-headed, 63 per cent compared to 37 per cent female-headed households' (Ghana Statistical Service, 1994:11). The same trend exists in the families of my respondents. Of the 160 cultivators who have families, divorced or widowed, 65 per cent were male-headed.

The size of a household partly influences one's decision to participate in urban agriculture. Ignoring socio-economic differences, the larger the household size, the more likely it is for members of that family to get involved in urban agriculture. This is because the larger the family, the more food it consumes. It does not follow that the larger the household, the larger the total income because some members of the family may not be in any income-generating employment. As Rakodi (1988) puts it, monthly income per capita declines with household size. It, however, follows that the poorer the household, the larger the size. The reason this occurs is that many poorer relatives move from rural areas to stay with relatives in towns and cities in an effort to secure jobs (see chapter 7). Also, 'poorer people [have] more children and [make] less use of fertility reduction measures, thus supporting the obverse side of demographic transition theory' (Robertson, 1984:223). Poorer people make less use of fertility measures because their 'dependence ... on their children is growing, a trend which may increase fertility' (Robertson, 1984:224). On the issue of urban agriculture, the larger the household, the less likely outside help or labour is used because there are more family members to help in farming.

The smallest household in this research consisted of only one person. The largest consisted of 14 members.

6.4 Socio-Economic Background

Contrary to general belief that urban agriculture is performed by the poor, the uneducated and the unemployed, recent findings show that urban farmers comprise a mix of socio-economic groups from various backgrounds. This assertion is supported by various researchers. For example, in 1994 Mougeot noted that, 'urban agriculture is not only the poor's business' (Mougeot, 1994:22). Urban farmers are both the rich and the poor; literates and illiterates. A survey of six Tanzanian cities by the Sokoine University of Agriculture revealed that animal breeding is a money-maker for top executives (Mougeot, 1994:22). In Dar es Salaam, urban cultivators are evenly distributed across educational levels (Sawio, 1993).

One of the most difficult kinds of information to solicit from Ghanaians is their income because they do not like to divulge their income to strangers. Consequently, in this work other variables like types of accommodation, educational background and information on whether one is employed and at what level, and whether a person is self-employed were used to determine the farmers' economic background. The assumption is that people with a high income will invest in their accommodation, they will be employed at higher positions as a result of their higher educational background or own businesses. Table 6.4 shows the educational background of the cultivators in this study.

The table confirms the assertion that people of all educational backgrounds are involved in urban agriculture. A further breakdown of the figures in the table showed that the educational standard of male farmers is generally higher than that of their females counterparts which tallies with national figures. In Ghana, the overall educational standard of males is higher than that of females. Statistics show that,

> Of the male population, more than 26 percent have never been to school, 33 percent have had only primary education, about 30 percent have attended middle school, and only 11 percent attended secondary school. Comparatively, 38 percent of the women have never been to school, 31 percent have only primary education and 25 percent have middle/JSS education. Only 6 percent have had more than 10 years of formal education (Ghana Statistics Service, 1994:14 & 15).

Males, especially husbands in Ghana have a higher educational standard than women because men marry women with lower socio-economic status than themselves.

Table 6.4 Educational Background of Urban Farmers in Accra

Level of Education	Frequency	%	Cumulative %
No Formal Education	49	24.5	24.5
Elementary	70	35.0	59.5
Secondary	34	17.0	76.5
College	13	6.5	83.0
University	34	17.0	100.0
Total	200	100.0	

Data gathered on the employment status of the farmers in this study also confirm that people of all backgrounds are involved in urban agriculture. See Table 6.5.

Table 6.5 Employment Status of Urban Farmers in the Accra Study

Status	Frequency	%
Formally Employed	86	43.0
Unemployed	72	36.0
Pensioned	7	3.5
Student	5	2.5
Self-employed	30	15.0
Total	200	100.0

Most of the farmers in the Accra study are formally employed. For this majority, by the tenets of the dependency model of Third World economies (see chapter 2), their income is not enough to sustain them so they have to seek extra income in urban agriculture. A sizeable number of the farmers (36 per cent) are otherwise formally unemployed. The reason for their involvement in urban agriculture is best explained by the labour-surplus model which stipulates that due to the fierce competition in the labour market there are limited available job positions. Many people are therefore not formally employed. The unemployed with no guaranteed income then farm in order to earn some income. In chapter 2, I showed that among informal sector jobs urban agriculture is the most probable activity the unemployed will engage in. This statement becomes more true for former junior Civil Servants. This is because experience in the Civil Service does not necessarily qualify a Civil Servant for alternative employment. In fact, his/her employability outside the Service decreases in inverse proportion to his/her length of service.

It should be emphasized that people with very high income are also involved in urban agriculture. These are people who do not have to farm to supplement their income, they are not weak in the labour market. Consequently, the conclusion is that they farm because, as the cultural lag model denotes, of their interest or previous knowledge in farming, a knowledge they acquired mostly from their parents in rural Ghana.

It is important to know the public's perceptions regarding the socio-economic status of people involved in urban agriculture because the former's perception influences their attitude toward the practice. This in turn influences government officials' attitude toward the practice (see chapter 5). Therefore I asked the respondents (farmers, non-farmers, officials) the following question, 'What category of people are involved in urban agriculture?' The answers I got from farmers for this question are tabulated in table 6.6.

The most important figure in the table is 104, which represents 52 per cent of the responses. According to 52 per cent of the farmers all types of people, thus rich and poor, are involved in urban agriculture. Sixty-seven farmers (33.5 per cent) said the poor form the majority of urban farmers. Like the conclusion drawn after examining the employment status of the farmers, this finding also negates one of the tenets of the surplus-labour model that urban agriculture is dominated by unschooled female labour, and the poor in general (see chapter 2).

It is mostly the poor farmers who claim that the poor form most of urban farmers. This is a rational response to the question because it is difficult for them as poor farmers to say that the majority of farmers are rich. At one point of this study, I asked some of them whether they would agree that middle/upper class people form the majority in urban agriculture. Here I quote one of them to summarize their response, 'How can I say that? Look at me, do I look like a rich man?'

The twelve officials and forty non-farmers in this study also mentioned that all types of people, particularly the rich, are involved in urban agriculture. This response is similar to the response found by other researchers in some cities of East African countries (Sawio, 1994; Bongole, 1988; Mtwewe, 1987 Mvena, 1986). Research in some East African cities confirmed the perceptions of respondents to the extent that one researcher declared, 'the assumption made that the low-income urban people are the ones engaged in farming does not hold. The high-income people are the majority in the business' (Mtwere, 1987:9).[3]

---

[3] This declaration is over-emphasized.

Table 6.6 Urban Farmers' Perception of the Types of People Involved in Urban Agriculture

Status of Cultivators	Frequency	%
Rich	10	5.0
Poor	67	33.5
Employed	4	2.0
Unemployed	12	6.0
Illiterates	2	1.0
All classes	104	52.0
Total	199	99.5

6.5 Length of Stay in Urban Area

Contrary to the proposition of modernization theorists, 'UA is not an occupation taken up by recent migrants' (Egziabher, 1994:91). Rather, urban agriculture is undertaken by established urban residents. I asked the 200 farmers in this study, 'How many years did you live in this city before you got involved in urban farming?' (See Table 6.7).

The table shows that 98 of the 200 farmers (49 per cent) lived in Accra six or more years before they got involved in urban agriculture. Sixty-three farmers (31.5 per cent) lived in the city 1 to 5 years before getting involved. It should be noted that some of the 63 farmers who lived in the city between 1 and 5 years before farming were on the borderline, i.e. they lived in the city almost 5 years before getting involved in farming. Included in the 'over 10 years' group are three people who stayed in the city almost 20 years before they got involved in agriculture. Nineteen farmers, representing 9.5 per cent of the total, were born in the city. Thus, 'Not Applicable' represents farmers born in Accra.

Various researchers have come to the same conclusion in that urban agriculture primarily involves people already established in urban areas rather than recent migrants. As early as 1958 in Pointe-Noire (Congo) it was found that the largest field area belonged to people who had been living in the city for 5 to 20 years (Vennetier, 1961:72).

Table 6.7 Number of Years Lived in Accra Prior to Farming

Years in Accra Prior to Cultivation	Frequency	%	Cumulative %
Less than 1 yr	20	10.0	10.0
1-5 years	63	31.5	41.5
6-10 years	45	22.5	64.0
Over 10 years	53	26.5	90.5
Not applicable	19	9.5	100.0
Total	200	100.0	

It is noted by another researcher that 'more than 60 per cent of Lusaka's urban farmers had been in the city more than 5 years before starting their plot gardens, nearly 45 per cent had not farmed for the first ten years' (Sanyal, 1986:15). Similarly, Tracaud's (1988:8) survey of 100 gardeners in Freetown and Ibadan showed that most urban residents live in the cities for years before starting cultivation. Lee-Smith and Lamba (1991) noted in a study in Kenya that urban farmers are not recent migrants. Drakakis-Smith notes that 'urban agriculture is not the accidental or temporary business of mostly recent immigrants from rural areas' (Drakakis-Smith, 1992:5).

Commenting on his own findings on urban farmers in Nairobi, Freeman (1991) stated that 'these figures are surprising, since they tend to refute one common hypothesis about urban cultivators—namely, that most are very recent migrants who have not yet attained the coveted urban wage jobs they ostensibly came to Nairobi to obtain' (Freeman, 1991:58). A study of urban farmers in Kampala revealed the average length of time respondents had stayed in Kampala was 25 years (Maxwell and Zziwa, 1992:29). These findings are further supported by Mougeot's assertion that 'city farming is not the business of recent migrants...' (Mougeot, 1993:4).

An important question that comes to mind is, if urban agriculture is a cultural practice as mentioned in the cultural lag model in chapter 2, then why do urban residents wait more than 5 years upon arrival to towns and cities before farming? In Accra, the farmers were asked individually, 'Why didn't you farm in this city until you have lived here for many years?.' The answers are summed up in two main reasons.

The first response can be summed up by Eziabher's assertion that 'people come to the city and gradually move on from their basic or most immediate situation through 'coping strategies' for their immediate problems' (Eziabher, 1994:91). Some of the newly arrived migrants did not engage in urban agriculture because they did not understand the land tenure system in the cities. They did not know who owned which land, the restrictions, and so forth. These are mostly people who did not have any strong existing networks in Accra before arrival. They underwent 'a transitional period during which they either established some degree of social and economic security or ...' (Dettwyler, 1985:247-248). In other words, they had to establish a social network to help them find available land. One of the farmers said,

> When you are new to a place you have to know the area before you do something like farming. It is like when you take a fowl to a new pen. It stands on only one leg for sometime till

> it becomes confident in itself before it stands on both legs. If you have not gained ground at a new place how can you perform confidently? How can you even get access to land when you're new to the city?

Secondly and more importantly, most of the cultivators in Accra came to the city to try to secure jobs in the formal sector, so they spent some time trying to satisfy that goal. According to them, agriculture was not part of the agenda or issues that brought them to the city. They got into farming when they had satisfied the issues, when they had given up solving the issues of formal job employment or when they found out that 'their initial adjustment to the first-stage possibilities are limited or unsatisfactory' (Egziabher, 1994:92).

The above assertions mean that wage employment is by far the most important need of new migrants to urban areas. One hundred and five farmers (52.5 per cent) came to Accra in search of salaried jobs or were transferred to the city in connection with their jobs. Therefore, the first stage of a typical new migrant's adjustment to urban life is searching for a job. Apart from wage employment, family re-union (24 per cent) is the most important reason people migrate to Accra, followed by education (10.5 per cent). Of the 200 farmers interviewed in Accra only three (1.5 per cent) said they came to the city for the purpose of farming. All of them came as children with their parents who came to farm, as they put it, 'several years ago.' All those who came to the urban area to farm reside at the periphery of the city where land is more plentiful.[4] When we ignore the group that came with their farmer-parents, none of the respondents came to Accra to farm. Farming is therefore not an important factor when one decides to migrate to an urban area. Not applicable (N/A) represents the response of the 19 farmers born in Accra. That only a few people born in the city are involved in urban agriculture is expected.[5] According to the cultural lag model of chapter 2, by virtue of their lack of rural background, people born in the city are not very interested in agriculture.

---

[4] Their farmer-parents are long dead.

[5] Some of the farmers born in Accra have at earlier periods lived in rural areas.

Table 6.8 Reasons for Migrating to Accra: Urban Farmers

Reason	Frequency	%
N/A	19	9.5
Employment[6]	105	52.5
Family	48	24.0
School	21	10.5
Farm	3	1.5
Other	4	2.0
Total	200	100.0

6.6 Origin of Urban Farmers

Most of the farmers in Accra were born in rural areas. This indicates that rural areas have important ties with urban areas.

Table 6.9 Birthplace of Urban Farmers

Birthplace	Frequency	%
N/A	1	0.5
Accra	19	9.5
Other town	72	36.0
Rural	108	54.0
Total	200	100.0

[6] Search for employment and transfer to Accra in connection with employment.

As reflected in the table above, 19 (9.5 per cent) of the farmers were born in Accra. Seventy-two farmers (36 per cent) were born in other towns/cities. This means in all, 45.5 per cent of the farmers were born in urban areas. It should however be mentioned that a large number of those born in urban areas had once lived in rural areas for employment purposes, or as youths for educational purposes. To confirm this assertion, I asked the farmers whether they have ever stayed in rural areas. The results are found in Table 6.10.

According to figures in the table, nine out of the 19 farmers born in Accra have rural experience. Similarly, 42 out of the 72 farmers (i.e. 58 per cent) born in other towns/cities have experience in rural life. In all, 140 out of the 200 farmers (representing 70 per cent) in this study have once lived in rural areas. In Table 6.9, it is shown that the majority (54 per cent) of the farmers in Accra were born in rural Ghana. This, and the fact that most of those born in urban areas have rural experience, also confirms the main tenet of the cultural lag model that urban farmers have rural background, that they farm partly because of their past background in rural agriculture. There are two other findings that support the cultural lag model. The first is that, according to the farmers, most of them learned firsthand agriculture from their parents who lived in rural Ghana. Secondly, most of the farmers have actually been involved in agriculture as school pupils in rural Ghana.

There was evidence in the field to negate the assertion that 'one-jump migration, directly from the farmstead to the capital, is the norm in Kenya and throughout Africa' (Freeman, 1991:58). I asked the farmers, whether they had ever stayed in other towns/cities prior to migrating to Accra. One hundred and fourteen farmers (57.5 per cent) answered in the affirmative. These people had either worked or attended school in other Ghanaian towns/cities, especially their regional capitals before, migrating to Accra. Only eighty-five farmers (42.5 per cent) said they moved directly from rural areas to Accra. It is interesting to note that 30 per cent of the farmers farmed in their previous urban areas. This means that before they started farming in Accra, the farmers had experience in either urban or rural agriculture, or both.

Table 6.10 Number of Urban Farmers Who Once Lived in Rural Areas

Birthplace	Lived in Rural Area			Row Total
	N/A	Yes	No	
N/A	0	0	1	1
Accra	0	9	10	19
Other town	1	42	29	72
Rural	0	89	19	108
Column Total	1	140	59	200

6.7 Logistics

Land is the most important factor when examining farming, thus a farmer should either own or rent the land he/she cultivates. In Ghana, land is either public property or freehold.[7] Generally, farming on land which is on freehold accords the farmer a more secured tenure than farming on public land. This is more so if the land belongs to the farmer or if he or she has an agreement with the landlord. Public lands are normally earmarked for various development projects so 'the land can be claimed at any time...' (Maxwell, 1994:61).

A typical urban farmer may cultivate on freehold land which belongs to himself/herself or another person. In the latter case, as mentioned above, land security is dependent upon the agreement of lease between the landlord and the farmer.

In Accra land may not be bought for the sole purpose of farming. When the government officials involved in this study were asked, 'Can a plot of land be purchased purposely for cultivation?' all of them answered, 'No.' A lease would not be issued if a prospective buyer indicated that the land would be used for agricultural purposes. Egziabher (1994) made the same observation in Uganda. She states that, 'Kampala City Council will not issue a lease if the proposed land use is agricultural' (Egziabher, 1994:58). One can however buy a plot for the construction of living accommodation or factory, including schools, hospitals and hotels. This is an indication that, formally, agriculture is not included in the planning of Accra and other Ghanaian towns and cities.

An official of the Agricultural Extension Services reported that even if it were permitted to purchase land for farming nobody would do that. His reason was that prices of land are so high in Ghanaian towns and cities that it would not be lucrative to buy plots for the purposes of farming. This indicates that the official does not consider urban agriculture lucrative enough to pay for itself. The purchasing of land is beyond the means of the average urban resident. As a result, many urban farmers in Accra do not own the land they cultivate (see Table 6.11).

---

[7] Individual or Communal Freehold.

Table 6.11 Land Ownership

Land Ownership	Frequency	%
Unknown	7	3.5
Self	37	18.5
Government	135	67.5
Other Individual	21	10.5
Total	200	100.0

Only 37 farmers (18.5 per cent) in this study farm on their own land. It should be mentioned that 33 of the 37 farmers who own the land they cultivate are enclosed or middle/upper class farmers (see Table 7.4 chapter 7). It follows that for open-space or lower class farmers in this study only 2.7 per cent own the land they cultivate. The low degree of land owned by urban farmers in Accra is similar to cities of other sub-Saharan African countries. For example, in Kenya only 41 per cent of urban cultivators own the land they cultivate. Similarly, only 33 per cent of the cultivators in Kampala own the land they cultivate (UNDP 1996).[8] Seven cultivators, representing 3.5 per cent of the respondents in this study, did not know the owners of the land they cultivated. Actually, this number would be higher if I took into consideration that many of the 67.5 per cent who identified the government as the owner of the land they were cultivating, did not know the government organ that owns them. They just know or think it is the government's property, and for most of them the government incorporates corporate bodies.

Though most of the open-space farmers claim to know the owners of public land they farm, they have not had any agreement with the former for the use of the land. Consequently, most farmers are squatters. They don't have any agreement for the use of public land because, as one of them put it,

> We don't know how to go about it. Who do I approach at the Electricity Corporation for an agreement to farm on their land?

---

[8] The high percentage of land ownership among urban cultivators in Kenya and Uganda may have been influenced by the fact that the researchers did not make any distinction between middle/upper class and lower class cultivators.

This quote of a cultivator summarizes the plight of urban farmers. This may indicate that the majority of urban farmers are willing to enter into a formal agreement with custodians of public lands. However, they don't know how to go about it. On the other hand, there remains a small number of farmers who hold a different view regarding the use of public land. The minority feel that public land belongs to the government thus, they should be able to use it without asking for permission. As one farmer said,

> The land belongs to the government and it is not being used so we have the right to use it without asking anybody. We are part of the government.

The farmers are reluctant to approach government officials for fear of complicating issues. According to them, they use the land without any immediate problems, so why should they approach officials for approval. They insist that government officials are corrupt, so if they approach them for approval they would demand money or crops from them. They are not willing to pay bribes to officials in order to use public land. Three farmers in this study insisted they pay rent to some officials for the use of government land.

Since the farmers do not enter into any agreement with government establishments and corporate owners of the land they cultivate, they are not informed when the land is needed for development. This in turn discourages the farmers from making long-term investment in their farms.

When farming privately owned land most farmers make arrangements with the owners. The first type of arrangement I will discuss is called *abusa*. This is a type of share-cropping agreement between a landlord and a tenant by which the tenant (here, the farmer) cultivates a piece of land without paying rent. However, when the cultivated crops are matured and harvested they are divided into three parts. The landlord takes two parts and the cultivator takes one part. *Abusa* is mainly practiced in the peripheral areas of Accra where people have purchased plots of land for the construction of houses. Prior to the house construction the land is leased to prospective cultivators. Of the twenty-one cultivators who farm other people's land (excluding government land), only 6 people or about 29 per cent farm under *abusa* share-cropping arrangements. The fact that the landlord takes two-thirds of the harvests indicates that the cultivator gains little in this type of arrangement. It is, therefore, not surprising that only a few farmers use this system of land tenureship, and most urban cultivators prefer to farm on public land where they do not pay any rent.

Another arrangement between private owners of land and urban farmers is called Care-taking. This arrangement is similar to *abusa* except that under this arrangement the farmer does not share his/her harvests with the landlord. In farming on a piece of land under this arrangement, the cultivator prevents others from encroaching on the land. In Ghanaian towns and cities, if an owner does not construct a house on his/her plot of land and it is allowed to grow into bush, another person may build on that land. Such incidences lead to litigations. To prevent other people from encroaching on their land, land owners who are not ready to develop their plots allow cultivators to farm the land prior to development. To farmers, this tenureship is better than *abusa* because they don't have to share their crops with their landlords.

Like rural farmers, urban farmers often hire the services of farm labourers. However, the labourers are hired for a shorter duration, mostly for land clearing. See Table 6.12 for an idea of the number of farmers in Accra hiring farm labourers.

Table 6.12 Number of Farmers that Hire Labour

Hire labour	Frequency	%
Yes	120	60
No	80	40
Total	200	100

Urban farmers in Accra mentioned three main reasons why they hire labourers for only specific assignments, and for only short periods. Firstly, the sizes of their farms are small. This is similar to other African cities, and since I could not measure the farms of the cultivators in this study I will use examples from cities in other African countries. For example, in Kampala, Maxwell and Zziwa (1992) noted that the size of areas cultivated ranged from as small as two meters by three meters to as large as ten hectares.[9] A survey in 1990/91 revealed that 64 per cent of the gardens in Dar es Salaam were less than 101 meters square, and 25 per cent under 51 meters square (Mougeot, 1994:19). Secondly, a significant number of urban farmers in this study (39.5 per cent made up the unemployed and the pensioned) are full-time farmers—as a result they have enough time to spend on their farming activities. Moreover, their farms are not in the bush as those of rural cultivators, so they spend long

[9] Only one person in their study cultivated an area of ten hectares.

periods of time in their farms. For example, while rural farmers have to leave their farms for home by 4pm (depending on the distance to home), urban farmers can be in their farms further into the evening because their farms are at shorter distances from their homes. Most cultivators spend over 40 hours per week on their farms. The farmers spend most of their time in their farms but, according to them, they do most of the farm work in the mornings and evenings.

The third reason why urban farmers do not hire labourers for long periods is that family members and other network members help on the farms especially during harvesting. Some of the farmers noted that whenever they needed some jobs done quickly they, asked their co-farmers to assist them.

6.8 Quitting Urban Agriculture

In chapter 2, I noted that urban agriculture is a permanent issue for those involved. To confirm this, I asked all the 200 farmers whether they would some day stop farming. The answers are tabulated below.

Table 6.13 Farmers' Intention to Stop Farming

Stop cultivating	Frequency	%
Yes	69	34.5
No	131	65.5
Total	200	100.0

According to figures in the above table, 65.5 per cent of the farmers do not intend to stop farming in the future. Sixty-nine farmers or 34.5 per cent said they will stop farming some day. The farmers were also asked to mention the conditions that could compel them to stop farming.

Table 6.14 Conditions Under Which Farmers Will Stop Farming

Conditions to Stop Farming	Frequency	%
Lack of market	1	0 .5
Loss of land	25	12.5
Sickness	154	77.0
Stealing	2	1.0
Other	18	9.0
Total	200	100.0

The most important reason for urban farmers to quit farming is sickness/death. I interpret this to mean that most of them will not stop farming as long as they are physically able. This is because the most important condition under which they will stop farming is beyond their control. Consequently, in essence none of the farmers in this study intends to stop farming. The next important reason is loss of farm land. Only one farmer, included in the 'other' category, mentioned that he would stop farming if he became wealthier. Apart from this individual, all the others indicated they will not stop farming even if they were offered better salaried employment. This finding answers an important issue raised in chapter 2—that workers will not abandon urban agriculture even if their income level becomes high enough to make food and other necessities easily affordable to them.

6.9 Summary

This chapter shows that people of all socio-economic status are involved in urban agriculture. It is also asserted that currently most farmers in Accra are men, despite the fact that during the colonial and early independence periods women formed the majority of urban farmers. Women are less involved in urban agriculture because they are not able to compete favourably with men for cultivable land. In addition, most Ghanaian women prefer retailing of commodities to urban agriculture.

Sixty-five per cent of the farmers in this study are married, 20 per cent have never married, and 15 per cent are divorced or widowed. Most of the farmers are between the ages of 35 and 44, and the larger their households, the more likely they are to engage in urban agriculture.

Most of the urban farmers in this study are well established in Accra meaning they are not recent migrants. Almost half (49 per cent) of the respondents lived in Accra for a period of six or more years before they farmed in the city. Most of them are originally from and/or once lived in rural Ghana.

The majority of the farmers do not have long-term investments in their farming activities. In addition, they hire farm labourers for only short periods. Some of those who farm other people's land make *abusa* and care-taking arrangements with the landlords. Data from the field indicate that for the farmers urban agriculture is a permanent issue. They say they will not quit farming even if their income from other sources improved.

From the issues raised in this chapter, one should be able to project those more likely to be involved in urban agriculture.

The location of farm reflects the status of the urban cultivators. This is studied in the next chapter.

7 Influence of Social Inequality on Urban Agriculture

7.1 Introduction

In the previous chapter, I discussed the general characteristics of urban farmers, ignoring the fact that in this study I identify two main types of farmers. These two main types of cultivators have some important differences, and to further enhance the understanding of the reader, I shall endeavour in this chapter to compare and contrast the two main types of urban farmers. This will be done on the basis of their access to related factors such as farmland, water, and education.

Social inequality is an important concept in human societies because it explains the distribution of resources. It denotes that some people have more access to resources than others. The effect of social inequality is greatly felt in urban agriculture. In sub-Saharan Africa, people of all social backgrounds are involved in urban agriculture. That means, to a degree, social inequality does not play any significant role as to whether a person will cultivate in an urban area or not. However, the effect of social inequality is remarkable when a person decides on the *type* of cultivation to be involved in. Thus, a person's social background, economic position or class determines what type of urban agriculture he or she chooses.

In sociological Marxist analysis, a class is a social group distinguished by certain common interests or characteristics and by the performance of certain functions or the holding of certain positions. It is impossible to specify all the classes into which a society is divided because there are too many 'varieties' of people or groups of people. Consequently, social scientists use various categories or characteristics when dividing societies into classes. According to Michelson (1976), people's homes, neighbourhoods, automobiles, speech, and so on convey class connotations. I add that in the case of urban cultivation the location of a person's farm is an indication of his/her social class. That means, as earlier indicated, social class conditions the location of farms. This is because people of different social classes have access to different levels of the living environment. Managers, professionals, and business people, for example, tend to live away from the core of towns and cities. They live in detached houses with plots of land surrounding them, which they put under cultivation. They cultivate areas relatively close to their dwellings.

Conversely, the working class and the poor in general tend to live in the core areas of towns and cities, and normally cultivate in the core areas. They live in houses with little or no land surrounding them. In cases where pieces of land surround their living places, the occupants normally do not have access to the land as it does not belong to them. As a result, they cannot farm it. The preceding observations indicate that social class is indeed relevant to the study of urban agriculture, especially to know who farms at specific places.

For the purpose of this work, only two categories of social classes are used—lower class, and middle/upper class. The categorization here is mainly on the basis of educational background, income, and the ownership of houses. Thus, all things being equal, one's educational standard determines his/her income. In turn, a person's income affects his/her choice of area of domicile, and this in turn affects the area of cultivation. It also means, as it will later become clearer, people from economically more developed regions have better access to education and resources than those from economically less developed regions. This means, in Ghana, ascriptive factors are very important in determining an individual's socio-economic position. It should be re-emphasized that this simple classification is for the purpose of this study only, and may not reflect the total class system in Ghana. The class system in Ghana deserves a separate study. For example, in the Ghanaian system old age is an important basis for classification. This is due to the fact that old people in a community may be the custodians of family or clan lands, which sets them apart from the others and gives them control over such lands.

Based on educational background, income, and the area of domicile, and consequently the location of farms, the concept of lower class is used to describe urban residents who cultivate land away from the immediate vicinities of their homes. I have termed this type of farming open-space farming/cultivation. A typical example of an open-space farmer is the roadside cultivator. Typically, the open-space cultivator is either an illiterate (or semi-literate), unskilled, formally unemployed, low-salary worker or lives on a low pension.[1] Comparatively, they are younger people who have recently entered the work force, or they may be low-level officials who have retired from active formal employment. On the average, an open-space farmer's household is large, containing six or more people. However, some of them may not have any spouse nor children. In addition, most of them neither own any piece of land in the urban area, nor do they own houses. They are, therefore, landless, and the majority are squatter farmers.[2] Most of them are men, and they cultivate mostly on a full-time basis. Of the 150 open-space cultivators interviewed, 81 representing 54 per cent were full-time cultivators.

On the other hand, the middle/upper class category of urban cultivators farm around their own or rented houses (homes). Their houses and consequently their farms are normally walled or fenced. As a result, I have termed their type of cultivation enclosed farming/cultivation. They are more likely to be either highly educated, successful business people, skilled, own the houses they live in, own plots of land in urban areas, highly salaried or live on a high pension. Table 7.1 shows the educational background of the cultivators involved in the Accra study.

As shown in the table, among the 150 open-space cultivators studied, 64 people or 42.7 per cent have only elementary education. Thirty-one cultivators or 20.6 per cent completed secondary and/or college education, and only 6 cultivators or 4 per cent have a university educational background. Forty-nine open-space cultivators, or 32.7 per cent have no schooling. This means that the majority (75.4 per cent) have no more than elementary school education.

On the other hand, among the 50 enclosed cultivators studied, 16 cultivators or 32 per cent have secondary/college education, 28 of them representing 56 per cent have university education. Only 6 enclosed cultivators or 12 per cent have only elementary education. This follows that most of them (88 per cent) have completed at least secondary education.

Basically, enclosed cultivators are property owners. They are middle aged and have been in formal employment for some time or well established in businesses of their own. Table 7.2 gives the employment status of both open-space and enclosed cultivators.

---

[1] There is normally a combination of these characteristics.

[2] Large-scale cultivators are not included in this analysis.

Table 7.1 Educational Background of Cultivators

Type of Cultivation	Educational Background					
	No Formal Education	Elementary Education	Secondary Education	College Education	University Education	Total
Open	49	64	28	3	6	150
Enclosed	0	6	6	10	28	50

Table 7.2 Employment Status of Urban Farmers by Type of Cultivation

Type of Cultivation	Employment Status of Cultivators					
	Employed	Unemployed	Pensioned	Student	Self-employed	Total
Open-Space	58	71	5	4	12	150
Enclosed	27	2	2	1	18	50

As Table 7.2 shows, 27 (54 per cent) of the enclosed cultivators in the Accra study are formally employed and 18 of them (36 per cent) established in their own businesses. The formal employment figure for open-space cultivators is only 38.7 per cent. Most open-space cultivators (47.3 per cent) are formally unemployed, and only 8 per cent of them run their own businesses. Comparatively, the unemployment figure for enclosed cultivators is only 4 per cent. The unemployed enclosed cultivators are wives of affluent people who have chosen 'to take care' of their children instead of seeking formal employment or running their own businesses.

The type of accommodation people have also reveals their economic position. By observing the differences between the cultivators' places of dwelling one is able to establish whether a cultivator is in the middle/upper or lower income group. Studying the data on their type of accommodation, I found that most of the open-space cultivators live in lower standard houses than do enclosed cultivators. According to information gathered on the field, all the enclosed cultivators of this study live in houses with a plot of land surrounding them, and most open-space cultivators live in houses that do not have extra land surrounding them. The few open-space cultivators living in houses with extra land surrounding them live in uncompleted houses of other people, as care-takers.

In the remainder of this chapter, I will compare and contrast open-space and enclosed cultivators in more detail. I will also study the advantages and disadvantages of open-space and enclosed cultivation. It will be suggested that the original home region of urban residents influences the types of cultivation they engage in. Thus, an urban resident who migrated from an area economically less developed is less likely to own land in the urban area of residence and more likely to engage in open-space cultivation, while an urban resident from an economically more developed area is more likely to own land in the urban area he/she is domiciled, and more likely to be involved in enclosed cultivation. Furthermore, the lower the socio-economic status of the cultivator, the more likely he/she will cultivate for sale. It would also be argued that geographical mobility affects the social mobility of urban residents differently. It is more likely for the social mobility of an urban resident who migrated from an economically developed area to rise than that of an urban resident who migrated from an economically less developed area.

7.2 Land Ownership

Land ownership is an important factor in urban cultivation because it, among other factors, influences the resources one puts into cultivation (see chapter 9). As a result, it affects the performance rate of cultivators. For the purpose of this chapter, I consider two hypotheses. The null hypothesis is, all types of urban cultivators have equal access to farm land. Therefore, access to farmland is unrelated to socio-economic background of urban cultivators. An alternate hypothesis is, urban cultivators do not have equal access to farm land. One of the questions I asked cultivators is, 'Who owns the land you cultivate?' Table 7.3 shows the answers.

Table 7.3 Land Ownership by Type of Cultivation (a)

Type of Cultivation	Land Ownership				Total
	Unknown	Own	Government/ Corporate	Other Individual	
Open-Space	7	4	122	17	150
Enclosed	0	33	13	4	50

As noted in the previous chapter, only 2.7 per cent of the respondents in the open-space category of farmers own the land they cultivate. One hundred and twenty-two open-space cultivators, representing 81.3 per cent, of the total number of open-space cultivators in this study farm on government/corporate land. Seventeen (11.3 per cent) of the 150 open-space cultivators cultivate land belonging to other individuals, and seven (4.7 per cent) did not know the owners of the land they cultivated. For the 50 enclosed cultivators, most of them cultivated land that belonged to them and/or land they held secured tenure over.

To find the chi-square, I collapsed the data in Table 7.3 into two categories: those who cultivated their own land, and those who cultivated government/corporate or other people's land. This shows that only four open-space cultivators owned the land they cultivated, and 146 cultivated other people's land. For enclosed cultivators, 33 owned the land they cultivated, and seventeen cultivated land that belonged to others (see Table 7.4).

Table 7.4 Land Ownership by Type of Cultivation (b)

Type of Cultivation	Land Ownership		
	Own	Other	Row Total
Open-space	4	146	150
Enclosed	33	17	50
Column Total	37	163	200

Chi-square = 99.74; df = 1; P = .01.

According to the list of chi-square values, the expected chi-square value with 1 degree of freedom is 6.635. My calculated chi-square is 99.74, meaning the two different types of cultivators do not have equal access to land. The null hypothesis is therefore rejected. Conditionally, the alternate hypothesis is accepted.

An indication from the above is that enclosed cultivators have more access to farm land than open-space cultivators do—but to what degree? To find this out I have converted the figures in the absolute frequencies in table 7.4 into proportions.

Table 7.5 Land Ownership in Proportions

	Land Ownership		
Type	Own	Other	Differences
Open	0.11	0.9	-0.79
Enclosed	0.89	0.1	0.79
Total	1	1	0

Proportions may be interpreted as probability, and probabilities vary between 0 and 1. Thus, they vary 'between certainty that a result will not occur and certainty that it will' (Hellevik, 1984:4). If none of the cultivators owns the land they cultivate the probability will be 0, and if all the cultivators own the land they cultivate the probability will be 1. Figures from the above table show that of all the cultivators who own the land they cultivate the proportion for open-space cultivators is only 0.11 (or 11 per cent). It also means that, the probability of an open-space cultivator in Accra owning the land he/she cultivates is just a little over a tenth of all cultivators who own the land they cultivate. Comparatively, there is a 0.89 (89 per cent) probability that an enclosed cultivator in Accra owns (or has a secured tenure over) the land he/she cultivates. It is also shown in Table 7.5 that of all the cultivators who farm others' land, there is 0.90 (or 90 per cent) probability that they are open-space cultivators, and only 0.10 (or 10 per cent) probability that they are enclosed cultivators.

To measure the degree of statistical association between land ownership and type of cultivation, I have calculated the difference in proportion between the two variables (i.e. land ownership and type of cultivation). The difference in proportion is 0.79, indicating that the probability of an urban resident becoming an open-space or enclosed cultivator is clearly different, depending on whether the cultivator in question owns land in Accra. The difference in proportion shows that there is a negative relationship between land ownership and open-space cultivation. What this means is that it is less likely (the negative sign) for an open-space cultivator in Accra to own the land he/she cultivates. It also means it is less likely for a person who does not own land to be an enclosed cultivator. The difference in proportion also shows that it is more likely (positive relationship) for an enclosed cultivator in Accra to own the land he/she cultivates. Land ownership is therefore an important determinant of the type of cultivation an individual engages in. Put figuratively, we have the causal relationship below.

Land Ownership ⟶ Type of Cultivation

Figure 7a Relationship Between Land Ownership and Type of Cultivation

So far, I have shown that urban cultivators in Accra do not have equal access to farm land. Enclosed cultivators have more access to land than open-space cultivators. An important question is, why do enclosed cultivators have more access to land than open-space cultivators? To answer this question I examine the home origins or regions of urban cultivators. My argument is that conditions prevailing in the home regions of urban cultivators determine their social background. This in turn influences the type of cultivation they engage themselves in. Thus, by observing the types of cultivation in Accra it is possible to hypothesize the economic conditions or infrastructural development of the home region of individual or groups of cultivators.

7.3 Home Regions of Urban Cultivators

I will start this section by studying the level of education in the ten regions of Ghana. See Table 7.6.

The northern area of Ghana consists of Northern, Upper West, and Upper East Regions. Figures from the above table show that 68.9 per cent of the people in the Northern Region of Ghana did not have access to formal education. The figure is 62.4 per cent and 58.7 per cent for the Upper West and Upper East Regions respectively. For the southern regions, the worst is Volta Region with 27.6 per cent of the people not attending any formal school. This figure compares very favourably with those from the three regions of the northern area. Similarly, the northern area does not fare well when it comes to access to secondary/higher education. For the Northern Region only 3.2 per cent of the population have access to secondary/higher education. For the Upper West Region, the figure is only 3.1 per cent of the population. With access to secondary/higher education in Upper East Region totalling 5.9 per cent of the population, it is better than the other two regions of the northern area. That is not all: on access to secondary/higher education the Upper East Region has a higher access than the Ashanti Region. However, this does not translate into a better economy in Upper East than in Ashanti. While the people of Ashanti Region can boost their income from cash crops like cocoa and kola, the people of Upper East do not have any such important cash crops.[3] Furthermore, the people of Ashanti Region make some money from prospecting minerals like diamond and gold. Consequently, while income from cash crops and minerals gives Ashantis capital to start their own businesses when they migrate to Accra, most Upper Easterners do not have any readily available capital when they migrate to Accra.

---

[3] Comparatively, the number of students from the Ashanti Region who further their education abroad is higher than Upper Easterners who study abroad. Indeed, talking about the number of students who further their education abroad, the Ashanti Region may top the list.

Table 7.6 Educational Level of Household Population (Ghana)

Region	Level of Education (%)				
	None	Primary	Middle/ JSS	Secondary/ Higher	Total
G. Accra	14.6	28.5	34.2	22.7	100
Eastern	22	34.1	35.2	8.7	100
Western	26	35	30.6	8.4	100
Volta	27.6	35.5	29.0	7.9	100
Central	27	34.4	32.4	6.2	100
B-Ahafo	26.3	35.8	31.8	6.1	100
U-East	58.7	25.8	9.6	5.9	100
Ashanti	25.2	36.1	32.9	5.8	100
Northern	68.9	21.1	6.8	3.2	100
U. West	62.4	25	9.5	3.1	100
Total (Ave.)	35.8	31.2	25.2	7.8	100

Source: Culled from Tables 2.6 and 2.7 in 'Ghana Demographic and Health Survey 1993' by Ghana Statistical Service (1994).

The consequences of unequal access to education are not difficult to predict. Without higher educational standards, people born and raised in the northern areas of Ghana do not secure any high salaried employment in their home regions and in Accra when they migrate. Similarly, since they do not have any viable cash crops and/or minerals they are not able to accumulate any significant capital before migration. Consequently, they are not able to operate their own businesses. Therefore compared to southerners, northerners do not have the appropriate skill and capital when they migrate to Accra. As a result, most migrants from the north end up in lower salaried jobs or are unemployed. The implication is that in Ghana, southerners are wealthier than northerners. This conclusion is supported by data provided by the Ghana Statistical Service (1995). Explaining some data, Ghana Statistical Service notes that 'the Northern, Upper West and Upper East regions are amongst the poorest regions, in terms of both incidence and depth of poverty. In general, these regions appear poorer than the others' (Ghana Statistical Service, 1995:17). Owing to their poorer conditions, most migrant northerners are not able to purchase land and houses of their own in Accra.

In order to give the reader an idea of where the cultivators in the Accra study hail from, I have divided Ghana into two main geographical areas—northern and southern Ghana.

Table 7.7 Area Origin of Urban Cultivators

	Origin of Cultivators in Accra		
Type of Cultivator	Northern	Southern	Total
Open-space	132	18	150
Enclosed	11	39	50

Figures in Table 7.7 show that of the 150 open-space cultivators in this study, 132 (88 per cent) come from the northern part of Ghana, and only 18 (12 per cent) from the south. Regarding the enclosed cultivators, 39 (78 per cent) come from southern Ghana, and only 11 (22 per cent) come from northern Ghana. Thus, northern residents in Accra who are involved in urban cultivation mostly practice open-space cultivation. They do not own the land they cultivate because the areas in which they were born and raised are economically less developed, and for that matter did not have the opportunity to acquire higher educational background. In addition, they might not have come to Accra with enough capital to start any viable businesses of their own.

It is obvious that among Accra residents a person's home-regional background influences his/her social status, and subsequently his/her chances of owning land in Accra. And as indicated in Figure 7a, land ownership influences type of cultivation. As causal effects, it is clear that home region takes precedence over land ownership because it is a person's area of origin that determines whether he/she will be able to own land in an urban area. Therefore, land ownership may act as an intervening variable in this cause-effect relationship (see Figure 7b).

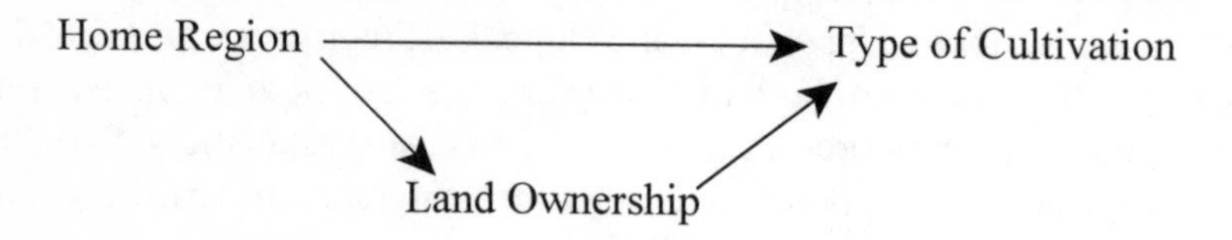

Figure 7b Relationship Between Home Region, Land Ownership, and Type of Cultivation

Taking all three variables into consideration, it becomes clear that land ownership acts both as a causal variable and an effect variable. It is a causal variable in relation to type of cultivation, and it is an effect variable in relation to home region. I offer a short explanation at the risk of repeating what has already been said. The arrow between home region and type of cultivation indicates that an urban resident's chances of being an open-space cultivator or enclosed cultivator is directly dependent upon whether he/she is from a developed or an undeveloped region of Ghana. The arrow between home region and land ownership shows that an urban resident's chances of owning land in an urban area is dependent upon whether he/she comes from a developed or an undeveloped area of Ghana, and this also in turn determines whether an urban resident will be an open-space or an enclosed cultivator (arrow between land ownership and type of cultivation). Thus, home region also has an indirect effect on type of cultivation through land ownership. The indirect effect means, for example, that not everybody from a developed area of Ghana becomes an enclosed cultivator: this is dependent upon the individual's ability to own land, and subsequently a house in Accra. It should, however, be remembered that people from more developed regions of Ghana have high probabilities of owning land in Accra.

The fact that most open-space cultivators are northerners is a confirmation that people from the northern part of Ghana living in Accra are comparatively poorer.

A related issue is the social mobility of urban residents. My proposition is that geographical mobility affects their social mobility. And it affects them differently depending on the home region of emigration.

7.4 Geographical and Social Mobility of Urban Residents

Despite general observations that many urban residents are unemployed or under-employed, geographical mobility (as noted in section 7.1) increases the social mobility or status of some individuals and groups of people, and it decreases the social mobility of other people. Comparatively, I expect people from the southern areas of Ghana to raise their social mobility when they migrate to Accra. Conversely, the social mobility of the people of the three northern areas of Ghana may not increase significantly when they migrate to Accra. As noted in the previous section, most people from the north do not have access to higher education, so they do not possess the necessary skills that may secure them well-salaried jobs either in the formal or informal sectors. Thus, they do not possess the skills that could secure them jobs which in turn would raise their social standing in Accra. In addition, as earlier indicated, they migrate to Accra without any substantial capital to start their own businesses.

An implication from the above is that Ghanaian northerners do not fare well in the competition for the limited salaried jobs in Accra and other Ghanaian towns and cities. In accordance with the tenet of the labour-surplus model, since most of the northerners are less educated or less qualified, they are rendered jobless. Furthermore, when they secure jobs they are more likely to be employed in low-paying ones, so they are not able to survive on their income. Therefore, those who are employed have to supplement their salary with income from other sources. The jobless have to generate income from the informal sector. As previously mentioned (in chapter 6), owing to their skills in agriculture or their agricultural background, they prefer urban agriculture to other informal sector activities. It is important to emphasize that most of the cultivators in this study who cultivate mostly for economic reasons are from this category of cultivators.

On the other hand, due to resources in their home regions, many migrants from regions in the south may come to Accra with some capital. With time, they start their own businesses and take advantage of the dense population in the city—they have a large market to operate their businesses. Setting up their own businesses increases their wealth considerably. In addition, due to their higher educational background people from the south are more able to secure higher-salaried jobs in Accra and other towns. Therefore, for people from southern Ghana geographical mobility translates into a higher social mobility, and they are more likely to cultivate due to their interest in agriculture and not necessarily for economic reasons.

The above confirms that geographical mobility affects people's social mobility differently, depending on their home regions. It also means, being born in the northern parts of Ghana is economically disadvantageous.

Having established that people from the northern areas of Ghana are more likely to engage in open-space cultivation, and those from the southern parts more likely to engage in enclosed cultivation, I compare and contrast the two types of cultivation in more detail.

7.5 Open-space Farming

Farming in open spaces—along gutter systems and motor roads, along hydro transmission lines, are some examples of open-space farming. As earlier mentioned, open-space farms are mostly found at the core of the city, though a considerable number of them are at the periphery.[4] Since the people mostly involved in this type of cultivation are not capable of purchasing land and buildings in Accra, they farm mostly on public land or land that does not belong to them.

In general, an open-space farm belongs to 'me.' It does not belong to 'us.' This is an indication that such farms belong to an individual, though other members of the family may occasionally assist in some aspects of cultivation. The insistence that an open-space farm belongs to 'me' suggests that the individual male farmer sees the income he generates from his farm as something he alone has control over. Therefore, when he sells the crops, the open-space cultivator alone decides what the money accrued from the sale is used for. The individual open-space cultivator in Accra is able to say the farm belongs to 'me' because he alone controls the technology, the participants,

[4] The division of a town/city into core and periphery is ambiguous in Ghanaian towns/cities because different areas of the towns/cities have their own core and peripheral areas. For example, Madina which is a suburb of Accra has its own core and periphery. However, I use core in this paper to indicate centre of main cities. That means the whole of Madina, for example, is a peripheral area.

and the goal of cultivation.

Technology is 'a mechanism for transforming inputs into outputs' (Scott, 1981:17). It consists in part of machines and equipments, technical knowledge and skills. Open-space cultivators were asked, 'who decides on the type of equipment used to cultivate your plot?' Every open-space cultivator said that they themselves decide on the type of equipment to be used on the farm. When further asked whether they conferred with others like family members before choosing the farm equipment they use, all open-space cultivators said they did not. As one put it,

> It is my farm. I work on the land most of the time so why should my wife or somebody else decide for me the type of equipment I should use?

The participants are the people who work on a particular farm. A typical open-space cultivator decides who helps him in the cultivation of his land. In addition, he decides how much time his family members put into the cultivation. He also decides which aspect of cultivation may require outside help. A frequent comment made by open-space cultivators during my fieldwork was, 'I decide who is to help me in cultivation. I know when I need help.' When I asked the cultivators, 'Can your spouse come to help you anytime he/she decides to?' Most of them said 'yes.' However, they emphasized that the final decision to receive assistance from their spouses and others rested on themselves, meaning they can refuse assistance from other family members if they decide to do so.

The final decision rests upon the individual open-space cultivator regarding the aim of cultivation. The concept of goal is used to mean 'conceptions of desired ends—conditions that participants attempt to effect through their performance of task activities' (Scott, 1981:16). The cultivator alone decides the crops to be grown, and whether the crops are for sale or for consumption. When I asked open-space cultivators if they conferred with their spouses regarding decisions on the types of crops they cultivate, all the married respondents said they alone made the decisions.

I also asked each cultivator to mention the most important factor that determines the crops he/she cultivates (see Table 7.8).

Table 7.8 Factors that Determine Choice of Crops

Type of Cultivation	Factors that determine crop choice (%)							
	Water	Demand	Taste	Storage	Fertility	Care	Maturity	Total
Open-space	18	32	8	0.7	22.6	0.7	18	150 (100)
Enclosed	6	0	62	0	14	18	0	50 (100)

The above table shows that class is an important factor when an urban resident decides on the types of crops to cultivate. As people who cultivate for sale, the most important factor influencing open-space cultivators in their decision on the crops to cultivate is the market demand. According to a cultivator,

> We cultivate purposely for sale so the most important factor, as far as we are concerned, is market for our crops. If we cannot sell, we will not cultivate.

Since they cultivate for sale, they choose crops that are high in demand and in short supply in order to attain maximum profit. In relation to market demand is the accessibility to market. As most customers buy directly at the farms, it is necessary for these cultivators to farm at places where their growing crops can be easily seen.

The second most important factor for open-space cultivators is land fertility. Here, the cultivators mentioned that the suitability of crops to the land they cultivate is vital. According to them, most of the seeds used to propagate exotic crops are imported. Consequently, apart from suitability, the availability and easy access to these seeds (not reflected in table) also play an important role when considering the suitability of crops to be grown. They also take into consideration the availability and affordability of fertilizers.

Next in importance are water supply and the maturity period of crops. The supply of water fluctuates according to the season, and this in turn affects the types of crops cultivators decide to grow. There is plentiful supply of water during the rainy season but not in the dry season. During the dry season cultivators, particularly those without a reliable supply of water, grow only crops that don't need much water. Some even abandon cultivation during this period of the year. Therefore, depending on the availability of water, crops grown during the rainy season may differ from those grown during the dry season.

The need to take the maturity periods of crops into consideration is necessitated by the fact that most open-space cultivators do not have any title to the land they cultivate. Therefore, they may be required to vacate their land with little or no notice.[5] Their crops may be destroyed at any time if the lands they cultivate are required for development. As a result, they grow crops that mature early.

---

[5] Most of the time they are not given any prior notice.

Only 8 per cent of open-space cultivators mentioned that their own taste determined the crops they cultivate. The low report of own taste as a determining factor is due to the fact that they do not cultivate for their own consumption. When I asked these 8 per cent to explain why it is so, they told me their taste is coincidentally similar to their customers'. An implication from the above is that the majority of open-space cultivators grow crops that they do not consume, i.e. crops that do not form part of their diet. Thus, they sell most of their harvests. To find out the percentage of the crops they sell, I asked them, 'What percentage of your crops do you sell?' The answers given are shown in the table below.

Table 7.9 Percentage of Crops Sold: Open-space Cultivators

Harvest sold	Frequency	%
Less than 50%	22	15
About 50%	18	12
More than 50% - 100%	110	73
Total	150	100

It is obvious from Table 7.9 that most (73 per cent) of the open-space cultivators in Accra sell most (more than 50 per cent to 100 per cent) of their harvest. Twelve per cent sell about half of their harvest, and 15 per cent sell less than half of their crops.

They harvest just enough at a time to satisfy the rate of demand of consumers. Consequently, storage and care needed especially for harvested crops are the least important factors open-space cultivators take into consideration when deciding upon the types of crops to cultivate.[6]

Open-space cultivation has some important disadvantages. First, it is easily prohibited. Formally, in Ghana, cultivating in open spaces in towns and cities is prohibited by law. A by-law, Act 359, states in part, 'no person shall grow crops at a place other than on land within premises ...' (Local Government Bulletin, 1972—see chapter 5). The Mazingira Institute (1987) made a similar observation in Kenya. It noted that the authorities prohibit

[6] As full-time cultivators, open-space cultivators have much time to devote to the care of their crops. Consequently, care needed by crops does not significantly affect their choice of crops.

cultivation by unauthorized persons of any unenclosed and unoccupied land in private ownership and of any government land and land reserved for any road. This means, open-space farming is easily prohibited because it is exposed or easily seen.

Second, crops growing on open-space farms are, as a cultivator said, an easy target for thieves because they are normally unprotected, especially at night. According to the cultivators, many open-space farms are along roads, so people passing by can easily harvest crops, especially non-root crops like maize, okra and pepper, and vegetables in general, without being detected. Thus, theft is a major problem that open-space cultivators have to contend with (see chapter 9).

Third, since they cultivate along gutters or open drainage systems their farms are at times unintentionally destroyed by public officials who desilt the gutters. This is how an open-space cultivator explained it,

> When these people come to desilt the gutters they throw the silt and garbage onto our crops, destroying them. Once, I pointed this out to one of the workers, and he asked me whether I expected them to take the garbage to their homes. He was very rude. I understand him to a point. They have to drain choking gutters but when they put the garbage on our farms, at the edges of the gutters, within a few days the garbage drops back into the gutters.

The fourth disadvantage of open-space cultivators, compared to enclosed farmers, is that most of them do not have a reliable source of wholesome water. The type of water used in cultivation depends on the type of cultivation: whether it is enclosed or open-space cultivation. The use of rain water is common to practitioners of both types of cultivation. However, unrecycled gutter water is the major source of water used by open-space cultivators (see the table below).

Table 7.10 Type of Water Used by Urban Cultivators in Accra

Type of Cultivation	Type of Water Used: Rain water	Tap water	Well water	Gutter	Other	Total
Open-space	46	24	14	63	3	150 (100)
Enclosed	10	33	7	0	0	50 (100)

Sixty-three (42 per cent) of the 150 open-space cultivators mentioned untreated gutter water as their most important water source. A few (14 people or 9.3 per cent) of them use well water. Twenty-four of the open-space cultivators, representing 16 per cent, use tap water. Tap water is available to some open-space cultivators, however, it is too expensive and therefore not an option for them to use on farms. Apart from the high cost, authorities do not approve of the use of tap water for cultivation. The tap water used by the few open-space cultivators are either public taps or corporate taps being used without the knowledge of the owners.

When asked whether their water supply is reliable, the majority of open-space cultivators answered in the affirmative, and most of them added that they farm all year round. This means gutter water is a very reliable source of water for open-space cultivators. The issue is whether unrecycled or untreated gutter water is wholesome. In chapter 5 (5.2), I mentioned the outbreak of cholera in Santiago (Chile) as a result of the use of gutter water.

Another disadvantage of open-space farming was identified by an open-space cultivator at Dzorwulu, a suburb of Accra. He said,

> Open-space cultivators who work along roadsides expose themselves to risks of being knocked down by motorized vehicles. Furthermore, they expose themselves to pollutants emitted from vehicles.

Generally, since they do not undergo periodic medical check-ups, the degree to which the exposure to pollutants affects them cannot be documented. When I asked the cultivators whether they suffered any illnesses connected to cultivation, only four said they had developed back problems as a result of their involvement in cultivation. I do not seek to give the impression that roadside cultivators do not suffer from exposure to pollutants. What I seek to indicate is that there should be a study on the effects of vehicular pollutants on roadside cultivators.

Open-space cultivators in Accra were asked to mention the advantages of farming along roads, and in the open. The most frequently mentioned advantage was closeness to the market. This is very important because they cultivate for sale. They use plots of land lying at the core of cities so they are close to consumers. Furthermore, since they cultivate along motor roads it is easier for consumers to purchase from them. During harvest time some cultivators attract buyers by putting up signs announcing that crops are on sale. Consequently, by farming along roads open-space cultivators take advantage of directly and/or indirectly advertising their farm products without spending money on advertisement.

Another advantage for cultivating in the open, mentioned by open-space cultivators, is that farming along roadsides makes it easier and cheaper for them to bring inputs like tools to their farms. And when they have to transport their harvests to the market the proximity of their farms to motor roads makes transportation easier.

Closeness to consumers or markets increases the profit of urban cultivators in another form. They avoid waste by virtue of the fact that cultivators can harvest and transport foodstuffs to the market within short notice, depending on the dynamics of the market. Where the market is close-by or where consumers come to the farm to buy, as mentioned earlier, cultivators harvest and sell in bits, as determined by market demands. If, on the other hand, the market is far away, the cultivators [as rational beings] harvest and transport a large quantity of foodstuff to the market at a time. Distributing in large quantities saves the cultivators both money and time commuting to and from the market. However, in an event of demand at that particular time dropping below supply, prices would fall, lowering profits. Furthermore, some of the harvested crops may not be sold and due to lack of refrigeration subsequently go bad, again reducing profit.

The economic behaviour of open-space cultivators partly determines their profit margin. They cultivate mostly exotic crops which attract higher prices than local crops that are produced more efficiently in the rural areas. They sell their crops and use some of the money realized to buy local foodstuffs. The difference between the money made from their sales and the amount spent in purchasing traditional foodstuffs becomes their profit.

Apart from the differences between sales and purchases as mentioned above, the profit of open-space cultivators is influenced by the total cost of their inputs. Thus, the traditional accounting phenomenon of higher income minus less expenditure is a source of profit. The most important is the price of fertilizers, tools and other supplies used in cultivation. The cost of transportation also affects the profit of urban cultivators.

As mentioned at the beginning of this chapter, occasionally other family members especially children and spouses help in cultivation. The cultivators were asked, 'For which farming activities do you hire labour or use other members of your family?' Those who have ever hired labour or used the services of family members mentioned that they use extra hands primarily for land clearing. Land clearing is perhaps the most tedious agricultural activity in areas where crude implements are used, so it is understandable that they use extra help. On a few occasions, they use the additional services of people other than family members during harvesting. This normally happens when they have to harvest a large quantity of crops within a short time. For example, matured vegetables should be harvested within a certain period of time.[7]

Most of the open-space farmers in Accra cultivate multiple plots at different locations. See Table 7.11 for the answers they gave when they were asked why they cultivate multiple plots.

Table 7.11 Reasons for Cultivating Multiple Plots

Reason	Frequency	%
Land Tenure	52	50
Theft	37	35.5
Diseases	9	9
Pests	6	5.5
Total	104	100

The most important reason given by the 104 cultivators for cultivating multiple plots is associated with land availability and land security. According to open-space cultivators, it is difficult to acquire an extensive area of land in the same place, therefore they have no other alternative but to cultivate in different areas. In addition, they cultivate multiple plots as

---

[7] Normally, the cultivators plant in stages in order to avoid all their crops maturing at the same time.

security against land reclamation by owners. As one cultivator said,

> If you cultivate at only one area and the owner comes and says that he wants his land back then you lose everything you have grown. You lose access to the only land you cultivate. However, if you have farms at different places and one is reclaimed you still have other areas to cultivate.

The next most important reason is theft of crops. The general notion is that if one cultivates at only one area and his/her crops are stolen, he/she stands to lose more because there would be no other crops to rely on.

One other reason is to guard against total crop failure due to crop diseases and pests. A cultivator near Nkrumah Circle said,

> No one wants to put all of his/her eggs in one basket. Our crops are constantly threatened by diseases and pests. Some of the crops we grow are new to us so we don't have much knowledge about the diseases and pests that attack them. Obviously, we know little about how to handle these diseases and pests if they should attack our crops. If you cultivate at only one place and the crops at that area are attacked by diseases then you lose everything.

Open-space cultivators do not do much in the form of processing because they harvest as demanded by consumers. Occasional processing before sale is done by middle-women (market women) who buy and sell urban cultivated crops.[8] The processing is usually limited to washing and cleaning of the harvested crops before they are sold in the market. Also, no major preservation is undertaken. Some preservation is done only when there is a left-over from the market, that is, when harvested crops taken to the market are not sold out. The left-over is sold the next day. According to the cultivators, though left-overs are rare, the lack of preservation methods has been the occasional source of wastage.

Answers from open-space cultivators show that the bulk of harvested urban cultivated crops are sold to people of higher socio-economic status, including many expatriates. Other important customers are hotels and restaurants who normally buy in bulk. The fact that most of their customers

---

[8] Some market women sell on commission basis.

are people of higher socio-economic status is expected because they cultivate crops which attract high prices—crops that are consumed mostly by the affluent in society.

When open-space cultivators were asked whether they need the services of Agricultural Extension Officers, almost all of them said 'yes.' They indicated that they need the services mostly in the form of advice for improved cultivation methods. However, as low status cultivators they do not have enough contact with such officials because, as one of them put it, 'we don't know or see them.'

7.6 Enclosed Farming

As previously mentioned, unlike open-space cultivators, enclosed farmers are mostly middle/upper class urban residents. They are mostly southerners, and are more likely to be found at the periphery of Accra and affluent locations at core areas. They farm on their own lands or lands leased to them. My data (see Table 7.3) indicate that 66 per cent of the enclosed cultivators in this study farm on their own land. It is also shown in Table 7.3 that 13 of the 50 enclosed cultivators studied farm on government or corporate land, and four farm on plots of land that belong to other individuals. It is worth mentioning that the enclosed cultivators who do not farm on their own land are top government or corporate officials who live in houses provided by their employers, or they have rented houses from individuals. For example, some university lecturers in this study live in university-provided accommodations, and the land surrounding the houses is accessible to them for cultivation. Consequently, unlike open-space cultivators, enclosed cultivators farm on land on which they have secured tenure. This indicates that none of the enclosed cultivators faces the threat of being ejected from the land they cultivate.

As earlier mentioned, contrasted with open-space cultivators, enclosed cultivators have smaller households and are mostly employed or run their own private businesses (see Table 7.2). In addition, as shown in Table 7.1, they are generally more educated than open-space cultivators. Thus, generally, they have access to more resources than open-space cultivators.

Within the middle/upper class category of farmers there are generally two main sub-groups. Group one comprises business people whose source of income is largely from trading and other such ventures. The majority of them are not highly educated. In addition, they are less likely to be directly involved in urban cultivation because a large portion of their time is spent on their businesses. It is not uncommon among members of this sub-group to hire the services of labourers for farming activities. Group two comprises professionals, managers, and other middle/upper class people with a higher educational background. Their wealth is generated from their salaries and in some cases from kick-backs. They, like members of the first sub-group, hire labour and use the help of other people in cultivation. However, compared to members in sub-group one, they are more likely to be directly involved in cultivation processes.

Enclosed cultivators' aim of farming is quite different from that of open-space cultivators. They 'farm to substitute healthier, home-grown food for store-bought products and for personal satisfaction from the act of cultivating' (UNDP, 1996:59). The following is what a woman cultivator said about the joy of growing crops:

> I enjoy seeing my crops grow. Then I say, yes I have turned an idle area into a productive one. My children ask me many questions about the crops growing in the garden, and I love to answer those questions. They learn a lot from gardening.

As earlier indicated, unlike open-space cultivators, enclosed cultivators farm mostly for home consumption, not for sale. They were asked, 'What percentage of your crops do you consume?' All of them said they consume all the animals and crops they produce. A few, however, mentioned that they give out some of their harvest to friends. Some researchers in this subject area mention that it is rather middle/upper class urban residents who cultivate for sale. However, it is obvious from this study that it is rather the low class who cultivate for sale. Generally, enclosed farmers are wealthier than open-space farmers, and the former cultivate for home consumption because they need fresh vegetables, not because they cannot afford to buy them. As observed by the UNDP 'the motives of middle- and upper-income home farmers are often nutritional (cleaner and healthier home-grown food for the family) and cultural rather than economic' (UNDP, 1996:53). Some cultivate out of the joy of seeing their crops grow. Thus, on the issue of production for sale/home consumption, there is a contradiction between my findings in Ghana and that of Maxwell and Zziwa (1992) in Uganda. They found that,

> If looked at on the basis of how much food is sold, it is the high- and middle-income households, with a few exceptions, which are engaged in the practice. The vast majority of low-

> income households sell or give away very little of what they produce (Maxwell and Zziwa, 1992:34).

Thus, Maxwell and Zziwa (1992) assert that middle and higher income people, rather than lower income people are engaged in production for sale. It is encouraging that after making such an assertion they went on to caution their readers that,

> There was only a total of five respondents that were classified as having a high income... this is clearly too small a sample from which to draw statistically meaningful conclusions about this particular group (Maxwell and Zziwa, 1992:34).

As mentioned earlier (see 'Open-space Farming'), cultivation for sale in Accra is done primarily by men from the lower class, mostly northerners. Similar observations have been made in Brazzaville (Congo). In that city it is observed that,

> Women from Makelele quarter of Brazzaville, Congo, some of them coming from affluent households, grow cassava, whilst poorer or unemployed men grow salad crops to sell to better-off households, hotels and restaurants (Binns, 1994:122).

Lower class urban residents are more pressed for the need to make money from agriculture to service their non-food needs. It should be remembered that in this study, 76 per cent of the open-space cultivators studied said they cultivated mainly for sale. Conversely, virtually all the 50 enclosed cultivators said they cultivated for home consumption and not for sale.

One important observation, among others, can be made from the discussion so far. That is, since in Accra men dominate open-space farming, men control the income generating part of urban agriculture, and women who dominate in enclosed farming produce mostly for consumption. It follows that the traditional division of labour in rural agriculture, where men control cash-crops and women control food-crops, is retained in urban agriculture.[9] This observation is true also for Brazzaville (Congo), and surely for other sub-Saharan African cities. On the basis of such an observation, it can be concluded that Ratta's (1993) suggestion that 'urban agriculture ... is a powerful tool to uplift women's social position' is untenable. Especially, in a lower class family (at least in Ghana) urban cultivation rather increases the economic and social power of the husband.

Just like open-space cultivators, I asked enclosed cultivators, 'What factors determine the crops you cultivate and the animals you rear?' The response (on crops) is shown in Table 7.8. Due to the fact that they cultivate for home consumption, the crops grown by enclosed cultivators are determined, first and foremost, by their own taste. Thus, 31 (62 per cent) of the 50 enclosed cultivators mentioned their own taste as the main determining factor. The next most important determining factor mentioned by the enclosed cultivators (18 per cent) is 'care of crops.' Since they are mostly full-time workers and 'busy' business people, enclosed cultivators do not have enough time to care for crops that require a lot of attention. This may explain why they do not cultivate many exotic crops. Land fertility (14 per cent) and water supply (6 per cent) are the other factors that determine their crops. The time crops take to mature, storage and market demand do not play any role when enclosed cultivators are determining the crops to cultivate. In the first place, they have secured land tenure so they are not threatened with eviction. Consequently, unlike open-space cultivators they are not worried about the length of time it takes for their crops to mature. As earlier noted they cultivate for home consumption, so they are not really affected by the market demand of crops. Like open-space cultivators, the storage capacity of crops is not an important concern of enclosed cultivators because they harvest crops in bits as needed.

An insignificant number of enclosed cultivators note market supply (note: not demand) and subsequently the price of crops as partly an influence on the crops they cultivate. According to these cultivators, though they cultivate for home consumption, they take the market prices of vegetables into consideration when deciding on what crops to cultivate. An enclosed cultivator noted,

---

[9] The distinction between cash-crop and food-crop is not based on differences between crops. It is rather based on whether at a particular time a crop in question is produced for sale or for consumption. Thus, a crop may be considered as a cash-crop at one period, and a food-crop at another period.

> If there is a plentiful supply of a particular vegetable on the market the prices go down. Actually, at certain periods of the year there is an overproduction of some vegetables like tomatoes. Due to the poor preservation facilities in the country most of the harvests go bad. Well, I plan my farming activities in such a way that I don't cultivate tomatoes during the glut period. I cultivate other vegetables and buy tomatoes very cheaply on the market. It is not because I buy them cheap but because then I can use my land for other vegetables.

In other words, crops that have a low market value may not attract the interest of some enclosed cultivators. On the other hand, if there is a high demand for a particular vegetable, the price goes up. In order to save money, some enclosed cultivators grow high-priced crops.[10]

Enclosed and open-space cultivators differ on the types of crops grown. As noted in the previous section, open-space cultivators grow mostly exotic crops. Enclosed cultivators on the other hand cultivate mostly traditional crops and traditional vegetables. Since the market prices of exotic crops are higher than those of traditional crops, I expect enclosed cultivators as rational beings to cultivate exotic crops rather than traditional crops. It is not very clear from the data at hand why, though, enclosed cultivators consume exotic crops most but cultivate less of it. However, as earlier mentioned it seems they are discouraged from exotic crop cultivation by the amount of work involved. Exotic crops need more attention in the form of watering, weeding, chemicals and fertilizer application. Due to the hot temperature, at times it becomes necessary to construct shades or sheds over the growing crops. This is not to indicate that enclosed cultivators are lazy, rather, as part-time cultivators they do not have enough time to pay special care to growing exotic crops.

Since cultivation is a part-time activity for enclosed cultivators, they cultivate in their spare time, especially early in the mornings and late in the evenings. Quite a number of this category of cultivators have servants or maids who help in cultivation. Just like in the case of open-space cultivators, the extra help is used mostly during land preparation activities, like land

---

[10] However, unlike open-space cultivators they grow a lot of traditional crops like cassava. Since the prices of traditional crops are cheaper this shows that their choice of crops is least determined by prices.

clearing and vegetable bed making. In addition, the extra help is used for watering of crops, and for harvesting. A female cultivator explains this vividly,

> I'm a busy lady but I want to produce some of my vegetables. My 'boy' helps me a lot. He prepares the land for planting... He also does most of the weeding. However, my husband and I do the sowing of seeds and the watering. I hate watering but I have to do it because the taps flow only at dawn and by then the boy is not yet in the house.

I asked enclosed cultivators, 'Who decides the types of crops grown. Is it the husband or the wife?' Eighty-five per cent of the respondents said the women normally decided what to grow. Ten per cent said it was a joint decision. Only 5 per cent said the men decided. When I asked the men why their wives decide on the crops they cultivate the most common answer was,

> It is the woman who decides our menu so she is the best person to decide the crops to cultivate. She knows what she needs to prepare our meals.

Where only one spouse was involved in enclosed cultivation it was always the wife. It should be noted that this is different from open-space cultivation, where it is more likely that the man of a family cultivates. Since it is more likely the wife who gets involved in enclosed cultivation, there are more female than male enclosed cultivators. This observation is not surprising because Africa's colonial legacy dictates that men work to earn cash, while women engage in subsistence activities. Rakodi (1988) lends support for this assertion when she notes that,

> In much of Africa, cultivation, especially for subsistence, is undertaken by women, although the tasks may be shared, and responsibility may be adopted by men if production is of cash crops and/or mechanised (Rakodi, 1988:496).

Where a couple is involved in enclosed cultivation, the woman and the children (where applicable) dedicate more cultivation time than the man. Information from the field shows that a woman spends approximately twice the amount of time her husband spends on cultivation. I found out from the cultivators that men spend less time cultivating because they take up family responsibilities or needs that are done outside the home. For example, if the family car needs to be taken to a repair shop, it is most likely to be the man who does that. In some cases, especially where the men feel their outside employment demands most of their energies, they assume only 'supervisory roles' in cultivation. As a lady cultivator explained,

> ... my husband doesn't really work in the garden. He hires labourers and makes sure the work is properly done.

Where a wife has a job outside the home, she may end up running triple shift—her job outside the home, house work and urban cultivation. Thus, irrespective of the demands of their outside work, women do not assume 'supervisory roles.' While men feel comfortable 'supervising' their wives, many women do not feel comfortable 'supervising' their husbands because a supervisory role at home may lead to conflicts between them and their husbands.

An enclosed farm normally belongs to 'us.' This is contrary to what exists in open-space farming where the farm belongs to the individual cultivator. This can be explained by the fact that middle/upper class families have closer-knit families. Also, unlike an open-space cultivator an individual enclosed farmer does not control or decide the participants or those who help in cultivating the land, nor does he/she alone decide on the technology or tools used, and the goal of cultivation. It is for the most part a joint decision within a family.

As middle/upper class people, enclosed farmers have greater access to inputs like hybrid seeds. Furthermore, due to their influential contact networks they have more access to specialists like Agricultural Extension Officers. It should, however, be noted that when asked whether they need the services of Extension Officers, only an insignificant number of them said yes. It is therefore an issue of those urban cultivators, who do not need the services of extension officers, having more access to them due to their resources.

Enclosed farming also has its advantages. First, since it is enclosed it is protected from thieves. It should be noted that theft of crops was one of the most important problems identified by open-space cultivators. Second, it is generally more difficult to prohibit, in that an official cannot walk into other people's enclosed compound without the owners' permission. As a matter of fact, by implication (as shown in Accra Metropolitan Assembly by-laws) enclosed farming is not prohibited in Ghana (see chapter 5).

Another advantage of enclosed farming is that it provides crops very close to cultivators/consumers. Thus, they don't have to transport their harvested crops from one place to another. Since the farms or gardens are close by, consumers harvest their crops as they mature, and as they need them.

Furthermore, since enclosed cultivation is undertaken by the consumers themselves, they make sure that only wholesome water is used. It should be mentioned that none of the enclosed cultivators in this study uses untreated gutter water. Also, only approved chemicals are used. This point is necessary because despite its known hazardous effect on humans, some cultivators in Ghana and other developing countries still use DDT (Gamalin 20) to control pests.

The most important disadvantage of enclosed farming was mentioned by a lady cultivator. She said it attracts snakes and other poisonous animals and insects to dwelling places. This is her experience,

> One evening I saw something like a rope on a pawpaw tree very close to our house. Out of curiosity I went closer to find out what it was. To my disbelief it was a snake. Oh my God, a snake.

Another disadvantage of enclosed cultivation is that it takes up spaces that could have been used as playgrounds by children. This is what a man said,

> My wife is very interested in gardening—perhaps, too interested in gardening. She has cultivated all the land around our house so there is not enough outer space for the children to run around. The children don't seem to care much about that. They always help their mother, they help us in gardening. When I told my wife that we should leave some space for the children to play she reminded me that they [children] play outside our compound. That is true but it is uncertain whether they play outside the compound because there isn't much space within the compound.

A disadvantage of enclosed cultivation is associated with water supply. As shown in Table 7.10, apart from rain water, enclosed cultivators use mostly tap water. However, when asked whether their water supply is reliable, most of them said no. In Ghana, tap water supply is not reliable, especially during the dry season. Consequently, inadequate water supply is a problem for some enclosed cultivators. Lack of water during the dry season compels some of them to suspend cultivation or limit their farming activities.

7.7 Summary

The effects of social inequality can be felt at different levels of the society. In this chapter, I showed that an urban resident's economic position determines the type of urban cultivation he/she involves him/herself in. I also indicated that in Ghana a person's class is significantly dependent upon the area of the country he/she was born and/or raised. Urban residents born and/or raised in the three northern regions of Ghana are more likely to be lower class, less likely to own land in urban Ghana, and consequently more likely to be involved in open-space cultivation. Due to their poorer economic situation, thus, due to the fact that they do not have other income or do not earn enough income from other sources, open-space cultivators cultivate mainly for sale.

On the other hand, urban residents born and/or raised in the seven other regions of Ghana are more likely to be middle/upper class, more likely to own land in urban Ghana, and more likely to be involved in enclosed cultivation. Unlike open-space cultivators, they cultivate mainly for home consumption. I established that for people born and/or raised in the three northern areas of Ghana, geographic mobility may not translate into higher social mobility. Conversely, for people born and/or raised in the southern areas of Ghana, geographic mobility is more likely to translate into higher social mobility.

In addition, the advantages and the disadvantages of open-space cultivation and enclosed cultivation were identified.

8 Effects of Social Networks on Urban Agriculture[1]

8.1 Introduction

An urban society is characterized by a combination of *Gemeinschaft* and *Gesellschaft* types of relationships. By this statement I mean for an individual, some forms of relationships are secondary, modern and loose or universalistic and other forms of relationships are involving, primary or particularistic and traditional. In a city, an individual will have universalistic relationships with the general population while, he/she may have a more particularistic relationship with members of his/her social networks. For many individuals, contact with the universalistic society is just a means of surviving in the city but for most people the particularistic society is where they feel at home. Thus, the interpersonal or primary network,

> provides the most fruitful micro-macro bridge. It is through these networks that small-scale interaction becomes translated into large-scale patterns and that these, in turn, feed back into small groups (Granovetter, 1973:1360).

Owing to the existence of primary networks, in a typical city in a developing country urbanization does not necessarily result in an anomic

---

[1] In this chapter, social network is synonymous to an informal group in the sense that the members maintain contact with each other and they eventually have a focal leader (or central figure).

dissolution of social relations. Thus, urbanization does not completely loosen social ties.

As social entities, human beings are dependent on each other. On several occasions people have to rely on the help of others in order to achieve their goals. As mentioned in chapter 2, new arrivals to urban areas have to establish social networks before they could successfully engage in farming. This indicates that establishing social networks is extremely important to new migrants because, among other factors, they can 'rarely put together the necessary access to land, water and other inputs' (UNDP, 1996:57). As noted elsewhere, 'social support from kith and kin is a principal way by which people and households get resources' (Wellman and Wortley, 1989:1). In Nairobi, for example, it is noted that men working in the city,

> Are embedded in a network of social relationships, a network which encompasses both the city and the rural home... In the process of establishing themselves in town, the men receive assistance from those already there (Curtis, 1995:137).

The types of assistance or resources available to an urban cultivator depend on the cultivator's community ties. Migrants in general and newly arrived migrants in particular expand their networks of assistance and increase the likelihood that they will at some point become givers of assistance as well as recipients. As explained by Blau, 'social exchange centers attention directly on the social process of give-and-take in people's relations ... ' (Blau, 1986:85).

In this chapter, I hypothesize that people prefer to farm near other members of their social network others they are familiar with. By using this coping mechanism, i.e. cultivating close to these people they engage in social exchanges beneficial to all involved. Further, I will use the theory which states that social exchange might lead to bonds of friendship (Wallace and Wolf, 1986:175) to show that the more farmers exchange materials and ideas, the more friendly they become toward each other. Thus, the greater the intensity of social exchange, the greater the trust it creates and the greater it integrates individuals into social groups.

In addition, it will be established that previous relationships play important roles in integrating newly arrived migrants into urban agriculture. I will also argue that people of the lower class are more willing to assist each other because they, as individuals, depend more on one another. Thus, the poorer the farmer, the more willing he/she is to assist other farmers. Through data collected, I will indicate that open-space cultivators are willing to form cooperative organizations but they would rather form the organizations with people they already know and trust.

I will also look at how change in the family structure of urban residents may affect their social network membership, and their access to kin labour.

There are several kinds of community ties, so an important issue to consider when studying social networks is the kind of community ties that provides one kind of assistance or another. This is studied in the next section.

8.2 Characteristics of Relationships

The characteristics of a relationship or the type of community ties determine the type and degree of assistance or support that is given and received. According to Wellman and Wortley (1989), supportive ties are a function of either: (a) the strength of the relationship, (b) the access the two persons have with each other, (c) the collective phenomena that affect interpersonal behaviour, (d) normative obligations between kin, (e) the characteristics and resources possessed by network members, or (f) the similarity or dissimilarity between the network members.

My focal point of discussion is based on the supportive ties mentioned above. The first to be considered is the strength of a relationship. In this book, it is measured by the intimacy of interactions. It is a matter of commonsense to observe that the more intimate a relationship, the stronger it is. For the purpose of this work, intimacy focuses on relationships between farmers who may be spouses, colleagues, or kin. Spouses are obliged to maintain relationships and to assist each other. Thus, the support provided is more or less involuntary. A fundamental question is, what type of support will be shown or given in an involuntary type of relationship of this kind? In a spousal relationship it is expected that the spouses will give all types of support to each other.

In urban agriculture, an intimate relationship should encourage spouses to jointly own a farm or garden and to farm together. In this study, I expect enclosed farmers (middle/upper class farmers) to be in more intimate relationships with their spouses. This is because the relationship between their families is more equalitarian (or less oppressive) than that of lower class families. As noted by Tepperman and Rosenberg (1995), in families of less-educated people there is still a lot of inequality between husbands and wives. I assert that greater equality in a family breeds intimacy, since none of the partners feels dominated. Consequently, I expect more enclosed farmers and their spouses to jointly own their gardens, and to have stronger relationships with their spouses than open-space farmers. In chapter 7, I mentioned that enclosed cultivators insisted their farms belong to their families, while open-space farms belong to individuals.

The second factor of interest is the access or contact two persons have with each other. In a relationship or social network, the more frequently the members interact or are in contact with each other, the more supportive the relationship, and subsequently the more mutual the relationship. This suggests that,

> Frequent contact encourages the provision of support by fostering shared values, increasing mutual awareness of needs and resources, mitigating feelings of loneliness, encouraging reciprocal rounds of support, and facilitating the delivery of aid (Hammer, 1983).

Proximity fosters frequent contact, awareness of problems, and faster and easier delivery of assistance. Consequently, people who farm close to each other have frequent contacts, and through discussions become aware of each other's problems. Similarly, since they farm close to each other, and have frequent contacts, they provide faster and easier assistance. In addition, frequent contacts lead to strong and supportive ties, and 'physical access aids the provision of both small and large services' (Wellman and Wortley, 1989:21).

Following from the above, I argue that since open-space cultivators cultivate near each other, they interact frequently. Consequently, compared to enclosed cultivators, as colleagues they have stronger and more supportive ties and are aware of each other's problems. They are more likely to discuss problems associated with agriculture, help each other in farming, and also give financial credit to each other. All these factors assist them in their farming activities.

Collective phenomena affect interpersonal behaviour, thus a network members' 'capacity to communicate, coordinate and control should increase the flow of support to its members' (Wellman and Wortley, 1989:3). For a network group to communicate, coordinate, and to control its members efficiently it should be binding, so if possible there should be an entity like a cooperative organization. I seek to impress that without organizational rules and written and unwritten regulations, a group may not be able to control its members efficiently. Furthermore, for proper coordination, a group needs a key figure or leader. Therefore, I propose that a group or network with a reliable central person or coordinator is more likely to attain better interpersonal relations, and subsequently better assist its members. Since most interactions between open-space cultivators are around their farming activities (thus, most of their network members are also urban farmers), they communicate, coordinate and control their members better on agricultural related concerns.

In addition, for a network group to be able to control its members it (the group or its members) should provide services that are vital to the individual members. Due to their low economic background and on the basis of rationality, some individual open-space cultivators are not able to obtain all the equipment they need for cultivation.[2] Consequently, they rely on their network members for such equipments. Therefore, my projection is that compared to enclosed farmers, open-space farmers are more restrained by their membership in network groups.

Kin members are also important sources of support. It is generally known that 'there are both cultural and structural pressures for kin to be supportive' (Wellman and Wortley, 1989:24). Consequently, all things being equal, people will receive more support from kin than from friends and other network members. In Ghana and other sub-Saharan African countries, the existing extended family-relation system widens kin support available to an individual. If a bond exists, whether active or not, the larger the number of a person's kin, the more resources he/she can gather if necessary. Unlike in North America and Europe where people 'have lesser expectations for supportive relations with extended kin' (Wellman and Wortley, 1989:25), in sub-Saharan Africa people have equal expectations for supportive relations with both extended and immediate kin. In some sub-Saharan African societies, people have more expectations for supportive relations with extended family than with immediate family. For example, in the Ashanti ethnic group of Ghana where inheritance is matrimonial, traditionally uncles are expected to cater for nieces and nephews instead of fathers catering for their sons and daughters. Since lower-class Ghanaians

[2] As rational beings, and due to their limited economic resources, some open-space cultivators do not purchase equipments they use only occasionally.

in urban areas are more willing to accommodate their kin (see Table 8.2), one can safely predict that open-space farmers in urban areas get more support in the form of credit, assistance in acquiring land and so on from kinship and network members than enclosed farmers get. For example, as shown in Table 8.1, 73 or approximately 49 per cent of open-space cultivators in this study acquired their farmland through the help of friends/relatives. None of the enclosed cultivators acquired farmland through such source.

Table 8.1 Sources of Farmland Acquisition by Type of Cultivators

Type of Cultivators	Means of Land Acquisition				
	Landlord	Friends/ relatives	Share- cropping	Other*	Total
Open-space	38	73	5	34	150
Enclosed	12	0	0	38	50

* Other includes own means and own land.

The next factor to consider concerns the characteristics and resources possessed by network members. By characteristics, I mean 'positional status that network members "possess" rather than qualities of their relationships' (Wellman and Wortley, 1989:30). Resources possessed by network members include capital, farm implements, high status, empathy and skill that the possessors may share with others. Since this is an exchange relationship, one of the most important issues is that most, if not everybody, in the relationship have something to loan or give out, and/or willing to give.

There should be a division of resources for network members to engage in an exchange relationship.[3] Also, 'people with high socio-economic status may experience frequent requests for instrumental support and companionship' (Dirk and Flap, 1988, quoted in Wellman and Wortley, 1989). Therefore, in a social network people with high socio-economic

[3] Here, I ignore socio-psychological reasons for exchange relationships.

status (here, those who have farm equipment which others require but do not have, those who possess vital information, and those who provide credit to others) will eventually emerge as the leaders or the most vital members of the network group.

Lastly, the similarity or dissimilarity between network members influences the amount and type of assistance or support they receive. On the similarity/dissimilarity notion there are two main arguments. According to similarity analysts like Lazarsfeld and Merton (1954), network members in similar structural positions tend to 'flock together' in strong 'hemophilic' friendships. Normally, network members have similar interests and characteristics like educational background and income levels, so they foster understanding and mutual support. Thus, for such analysts people with similar characteristics and status provide the best support for each other. In this type of relationship people may provide support and other services not because of reciprocity but out of the joy of helping network members.

Dissimilarity analysts on the other hand contend that it is mostly network members who occupy interdependent structural positions who exchange support. According to these analysts, division of labour [and division of resources] which cuts across social categories fosters solidarity and satisfies mutual needs (see Blau and Schwartz, 1984). The contention is that in a network comprising socially and economically dissimilar people the members have different access to resources, so they depend on each other.

My position as far as the similarity/dissimilarity argument is concerned is a complementary one. In the first place, normally people with similar backgrounds become members of the same social network. Thus, people with similar educational background and similar social status belong to the same social network group. In this similar-membership network group there is dissimilarity in the sense that some people have more access to resources than others. Apart from that, people in a network possess different materials and information, so depend on each other. For example, considering their educational background and income the cultivators at Dzorwulu are similar. However, apart from the basic farm equipment like hoes, they possess different equipment and have different access to information. This makes it necessary for them to provide assistance or support to each other.

In the next sections, I will demonstrate how the supportive ties discussed above apply to the cultivators in this study.

8.3 Cultivation Close to Each Other

According to a school of thought in the tradition of exchange theory as mentioned above,

> People tend to associate disproportionately with others proximate to them in social space, that is, with others who belong to the same group or whose social status is close to their own, whatever the dimension under consideration (Blau, 1986:93).

In respect to urban agriculture, following from the above hypothesis, it can be said that similarities in education, work, ethnic background and social class enhance the likelihood of urban residents farming close to each other.

My fieldwork supports the above theory that people of the same or similar social status associate with each other. I observed that people of the same ethnic background and educational background tend to cultivate near each other.[4] A newcomer to Accra, indeed to an urban centre, usually chooses to stay among people he/she shares a similar language and culture with. One of the most important reasons for this occupance is that in choosing friends people normally seek those who are similar to them. I asked open-space cultivators why people who cultivate in the same area spoke the same language and were similar in other respects, i.e. why people of the same ethnic and socio-economic background congregated in the same farming area. The most important reason given was that, they knew and understood each other so cooperated better.

Another important determinant is the way rural people migrate into urban areas. The high increase in the population of urban areas experienced in sub-Saharan African countries is first and foremost due to migration from rural areas (see Oucho and Gould, 1993). New arrivals may come on the invitation of people they have known in their villages in rural areas. Even when not invited, upon arrival to urban areas new migrants tend to seek accommodation or shelter from people they know from their villages. This behaviour is explained by the chain migration theory which states that when people migrate to a new area, they relocate at areas where similar, often

---

[4] This is most true for open-space cultivators. See chapter 7.

related, people are already living (Tepperman and Rosenberg, 1995:169). Thus, the complexities of urban life to the newcomer makes it necessary for him or her to seek initial assistance.

Following from the above, it should be noted that moving to cities is not only a spatial phenomenon, involving a reduction of certain distances to various points of economic attraction, but a social move aimed at integrating someone into a social network (Gregory et al, 1985) The social network may be provided by a family, friends, ethnic or political affiliations that eventually give access to job opportunities and capital and material borrowing. When asked, 'Did you know anybody in Accra before coming here?' 85 per cent of the cultivators answered they had some colleagues in the city before migrating there. Further probing revealed that most of the people they knew were originally from the same villages as themselves. Thus, 'in order to make contacts in the urban area, the individual makes use of rurally-based relationships' (Curtis, 1995:138). Due to these types of relationships, entry into the informal sector of the urban economy (including urban agriculture) 'is characterized by relatively easy entry because of the help rendered by those who are part of it to newcomers from the same village or locality' (Ibnoaf, 1987:10).

The above observations underline the importance of kinship, friends and other network members in drawing rural residents into urban areas, and in helping them get involved in urban agriculture. An important question that emerges out of the above observation is, if social network members assist newcomers as shown in Table 8.1, then why does it take the new arrivals approximately five years before they get involved in urban agriculture? As mentioned in chapter 6, new arrivals to urban areas come with the intention of securing salaried jobs. Consequently, they spend their first years in search of such jobs. Furthermore, in most cases established urban resident colleagues and kin use the services of the new migrants (see section 8.4). Thus, a combination of time spent looking for salaried jobs, and their social network members using their services, delay their entry into urban agriculture. In addition, a significant number of them do not have the necessary skills that will land them high-salaried jobs. Therefore, even after securing salaried jobs they still have to engage in urban agriculture to supplement their income.

8.4 Influence of Socio-Economic Status on Social Networks

The socio-economic status of an individual determines the networks he/she joins in that, as mentioned in the previous section, people of the same economic background and characteristics normally stick together.

The social networks of middle/upper class people are different from those of lower class people. Lower class people form their own networks, as middle/upper class people form theirs they hardly mix together in a network. The most important reason seems to be that these two groups of people have different interests. Accordingly, similarities in background, experiences, and social position make it likely that people exchange mutual support for their opinions and conduct, furnishing incentives for social interaction. As noted elsewhere, ingroup preferences 'induce persons to associate with members of their own group themselves, and they also induce them to approve of others who choose ingroup associates and to disapprove of those who choose outsiders' (Blau, 1986:97).[5]

In Accra I asked the respondents to list their friends or people in their social networks. In addition, they were asked to mention their friends' social class. Social class was measured using the following variables: living place, educational background, place and position of employment, and income. The data collected on this issue show that 98 per cent of the lower class respondents had all their friends in the lower class bracket. The two per cent that said they have some friends in the middle/upper social class bracket only worked for the so-called middle/upper class friends. When these people were asked whether their middle/upper class friends also considered them friends, they said they thought so. It seems the so-called middle/upper class friends use these people for menial jobs, and it is these calls to work that they (lower class) interpret as friendship. Similarly, all the friends of the middle/upper class respondents in this study are also middle/upper class.

It is not surprising that lower class and middle/upper class people do not associate with each other very much since it is known that 'the greater our ingroup bias in some respects, the more we restrict our choices in others, increasing the constraints to maintain intergroup relations in these other respects' (Blau, 1986:95). Middle/upper class people are suspicious of the lower class. Indeed, 'members of all classes tend to hold generally

[5] It should be remembered that dissimilarities also induce members of a network to exchange assistance or support.

negative images of the lower class...' (Tepperman and Rosenberg, 1995:132). After all, the middle/upper class determine deviant behaviour, and accordingly they term non-middle/upper class conformity deviant behaviours. In the eyes of these people, most deviants in the society are people in the lower class category. On the other hand, the lower class are biased against the middle/upper class. The former suspect that members of the latter groups exploit them.

Ghanaian middle/upper class people are less willing to accommodate their kin who migrate from rural areas, though they may assist them by giving them some money. They are unable to accommodate new migrants because of, among other factors, the nature of the houses they live in. The houses are self-contained and not suitable for visitors.[6] One of the questions I asked urban cultivators in this study was, 'Are you willing to accommodate migrants from your village/town even if they do not inform you before coming?' Most of the middle/upper class cultivators said they would not be able to accommodate new migrants (see Table 8.2).

Table 8.2 Urban Cultivators and Assistance to New Migrants

Type of Cultivators	Willingness to Accomodate New Migrants		
	Yes	No	Total
Open-space	128	22	150
Enclosed	4	46	50

The above table shows that only four enclosed cultivators, i.e. 8 per cent, of the middle/upper class cultivators in this study were willing to accommodate new migrants. As one of them explained,

> I'm aware of the so-called African hospitality. However, I live in a self-contained house with my wife and children. We have enough rooms for ourselves but we cannot

[6] Unlike a traditional house, a "self-contained" house contains few rooms (1-3 or 4 rooms), with no open space in the middle of the house. It is a more closely-knit type of house.

> accommodate others especially people who come unannounced. We are not used to living under crowded conditions, and we don't want to make our house a haven for new migrants. Migration to the cities should be stopped anyway.

Other researchers also note that middle/upper class people are less accommodating of their kin. For example, Caldwell (1968) makes a similar comment concerning Ghanaian urban elites. Oke (1986) also notes that in Nigeria socially upwardly mobile people attempt to avoid their kinship responsibilities.

Lower class people on the other hand have broader social networks, i.e. the number of people involved in their networks is large. They live in compound houses where many different families share the same kitchen, washrooms, and the like. Therefore, they interact more with other people. Furthermore, lower class urban residents are more willing to accommodate new migrants. When asked whether they were willing to accommodate migrants from their villages/towns even if they did not inform them before coming, 85 per cent of the open-space cultivators in this study answered in the affirmative. This is very much the reverse of the response given by middle/upper class urban cultivators. Lower class cultivators are more willing to accommodate new migrants because, as a lower class cultivator put it,

> Since we got settled in urban areas through the help of others we feel obliged to assist other people who ask for our assistance. It is an issue of give and take.

Another factor that induces lower class urban farmers to accommodate new migrants is their low income. They have lower income so they deem it necessary to eventually pull material and human resources together with new migrants in order to make ends meet. As earlier noted, some of them also exploit the labour of newly arrived migrants. A newly arrived migrant who helped his uncle on the latter's farm at the time of this study said this,

> I just arrived in the city [Accra], and since my arrival my uncle has been accommodating me. Actually, he assists me with everything I need. I don't pay him anything. However, I go on errands for him, and help him in his

> farm. I do all sorts of work for him. Our relationship is symbiotic, and it will be like this till I move away on my own.

Moreover, the networks of middle/upper class urban farmers are looser than those of lower class cultivators. This looseness is due to, among other factors, the fact that they do not exchange much in the form of tools, information, credits, and the like. Middle/upper class people, in this case enclosed cultivators, can afford all the tools they need for cultivation and they gather most of the information they need on agriculture from books, so they do not seek much information from others. On knowledge or information on improved agricultural practices most of them said they either read it from books or had acquired the knowledge while they were in Secondary Schools, especially where Agricultural Science was taught as a subject. They also acquire a lot of their information from agricultural specialists.

I found there is a great deal of exchange of gifts and information among members of networks of lower class urban cultivators. Owing to the high degree of interaction among them, the social networks of lower class farmers are more binding or involving than those of middle/upper class farmers. This supports the theory which states that the more cultivators exchange materials and ideas, the more friendly they become toward each other.

8.5 Reachability, Multiplexity, and Intensity of Social Networks

According to Mitchell (1987), there are three network concepts which underpin the classic theses of urbanism. These are reachability, multiplexity and intensity.

Reachability is used to mean 'the extent to which links radiating out from some given starting person through other persons eventually return to that same person' (Mitchell, 1987:304). This measures 'how true is it that everyone knows everything about everyone else?' (Rogers and Vertovec, 1995:16), and how smooth and fast information travels through a group of people. Reachability increases over time because the higher the frequency of interaction between people the more familiar they become. In Section 8.3 I noted that lower class urban farmers interacted more frequently with people in their social networks than middle/upper class urban farmers do.

Therefore it follows that all things being equal, lower class urban cultivators share information more than middle/upper class urban cultivators..

Urban residents who cultivate in the same geographical area interact a lot through the exchange of tools and sharing of farming techniques. Therefore, familiarity and consequently trust is very important in a network group. At Dzoworlu area of Accra, for example, the cultivators take their lunch break at the same time and actually dine from the same bowls in groups. Every cultivator knows much about other cultivators, and their friendship or interaction is extended beyond the farming areas. Social exchange processes among the open-space cultivators have led to social solidarity among them. As one of them noted,

> When any one of us encounters problems we rally behind that person. It is not only problems encountered at the farming site but all types of problems. When a colleague lost his son two years ago we, the farmers over here, pulled resources together to help him with the funeral.

In Accra, the tighter relationship between lower-class cultivators was further shown through the snowball effect of this research, albeit on a small scale. Some open-space cultivators recommended their colleagues for interviewing.[7] Multiplexity 'indicates the extent to which two persons are linked in more than one way' (Rogers and Vertovec, 1995:17). As earlier mentioned, the linkages between the cultivators in Accra do not end on the farm, they extend beyond. Middle/upper class urban cultivators are linked in many more ways than are lower class cultivators. For example, members of a network comprising middle/upper class people may also be members of various social organizations, members of the same political organization, they may serve as members on the same board, and they may be colleagues at work. Urban cultivators in Accra were asked, 'How many non-credit organizations do you belong to?' After that they were asked, 'Are your friends members of all the organizations you belong to?' The answers given for the first question indicate that middle/upper class cultivators belong to many more organizations than lower class cultivators do.

---

[7] Only one enclosed cultivator recommended a friend for interviewing.

Table 8.3 Cultivators' Membership in Non-Credit Organizations

Type of Cultivation	No. of Organizations			Total
	None	1-3	4+	
Open-space	111	36	3	150
Enclosed	2	10	38	50

Table 8.3 shows that 38 (76 per cent) of the enclosed or the middle/upper class cultivators belong to four or more organizations. Only two (4 per cent) of them do not belong to any organization. Comparatively, 111 (74 per cent) of the open-space or lower class cultivators are not members of any organization. Thirty-six (24 per cent) of them belong to only one, two or three organizations. Only 3 (2 per cent) of them belong to four or more organizations. The middle/upper class cultivators mentioned that most of their friends or network members were members of most of the organizations they belong to. Lower class farmers said the same.

Intensity is 'the degree to which individuals are prepared to honour obligations, or feel free to exercise the rights implied in their link to some other person' (Mitchell, 1987:27). To honour obligations may have different connotations. For the purpose of this work, it is used to mean reciprocity of gifts, the sharing of information and helping other network members to acquire farm necessities like cultivable land and tools.

As shown in Table 8.1, most low class cultivators get access to cultivable urban land through the help of relatives and/or friends. Education is an important factor in determining the status of people. Figures in Table 8.4 confirm that lower class cultivators, here, people with lower educational levels, are more likely to seek help from others like friends and kin in acquiring farmland. In the Accra study, 49 per cent of the cultivators who have no formal education and 50 per cent of those with only elementary education sought the assistance of friends and kin in acquiring farmland.

On the other hand, only 32 per cent of those with secondary school education and 23 per cent of those with college education got their farmland through friends or relatives. It should be noted that none of the cultivators with university education sought the help of others in getting access to farmland. Thus, most enclosed cultivators and people with higher levels of educational background do not need the assistance of friends and kin in acquiring farmland.

The above discussion emphasizes that intensity is greater among open-space cultivators than among enclosed cultivators. Social networks are more important for open-space cultivators because as individuals most of them lack the material resources to acquire all the necessary information they need. Some of them cannot read and write, and many of those who are literate do not have access to books which can provide them with improved agricultural practices, so they exchange information a lot.

Table 8.4 Sources of Farmland Acquisition by Level of Education

Level of Education	Means of Land Acquisition				
	Landlord	Friends/ relatives	Share-cropping	Other*	Total
No Formal Education	8	24	2	15	49
Elementary	14	35	0	21	70
Secondary	13	11	3	7	34
College	3	3	0	7	13
University	12	0	0	22	34

* "Other" includes own means and own land.

8.6 The Importance of Social Exchange

The willingness to share is very vital for urban farmers because it makes available to individual farmers tools and other necessities they could otherwise not have afforded. The issue is, 'Under which condition do urban farmers exchange farm equipment?' According to Ekeh (1974), a person will enter into a social exchange transaction with others only if he/she believes the exchange transaction will bring success. Urban farmers will lend out their tools and other farm equipment only if they know they will in the near future borrow from others in their social networks: that is, if they know they will benefit from the exchange. Though every cultivator in Accra owns almost every piece of equipment they use on their farms, occasionally they borrow some tools from their colleagues. These are normally tools they do not use frequently. For example, a cultivator who does not use a hand spade very frequently does not buy one. Rather, he/she borrows from his/her colleagues.

Similarly, some urban cultivators in Accra share crops. Comparatively, open-space cultivators share crops more than enclosed cultivators do. Urban cultivators in Accra were asked, 'Do you give some of your harvests out as gifts?' The answers received from the cultivators are shown in Table 8.5.

Table 8.5 Number of Cultivators Who Give Out Some Harvest as Gift

Type of Cultivation	Crops Given Out as Gift		
	Yes	No	Total
Open-space	114	36	150
Enclosed	10	40	50

Figures in Table 8.5 show that 114 (76 per cent) of the open-space cultivators in this study give out some of their crops as gifts. A significant proportion was mostly given to other cultivators and network members signifying that the act was symbolic. On the other hand, only few (20 per cent) enclosed cultivators give out some of their crops as gifts. In fact, that open-space cultivators share their harvests more than enclosed cultivators is expected. In the first place, the social network members of open-space cultivators are poorer and depend on each other for assistance. They are therefore willing to exchange. Secondly, as previously noted, the social networks of lower class farmers involve more interaction than those of middle/upper class farmers. Lower class cultivators interact more frequently in the field during cultivation. As earlier mentioned, most of them rest together from their work. That is, when they take rest they congregate and even eat together. The above discussion supports the assumption that, all things being equal, the more frequently people interact, the more supportive their relationship.

Among the open-space cultivators in Accra, the exchange of information and/or tools need not be reciprocal. All of them mentioned that they were willing to lend their tools to their colleagues irrespective of whether they would one day borrow from those who they lend to. As an open-space cultivator put it,

> It does not matter whether I borrow things from the person I lend to. He might not have any tool that I'll need but if he needs mine he can come for it. For me, the most important thing is that if I need help he'll be willing to help though he might not be able to. It is the willingness that matters.

So, in this exchange game of the cultivators the most important factor is the knowledge that everybody is willing to share important information or equipment with others. This type of exchange among the cultivators is not influenced by the commodity involved but by the act of giving and receiving. Therefore, the motives of the exchange are social psychological. Due to the exchange of goods and services among the cultivators, the average cultivator is integrated into the social group in which he/she is a member. To the question, 'Who would you borrow farm tools from?' all of the open-space cultivators answered they would borrow from anybody farming in their area of cultivation. This implies they are more comfortable borrowing from their social network members.

Many urban farmers operate on small-scale farming, and/or farm on plots of land not belonging to them. Consequently, they are not able to pledge their farms for loans from formal financial institutions. As a result, they have to look elsewhere to secure loans or capital in order to improve upon their farming operations. Information from the field indicates that the capital comes from two main sources:

1. A farmer may borrow money from a family member, a friend or other individuals in their social networks. It should be mentioned that, depending on the lender, this method of money borrowing may attract an exorbitant interest rate.
2. A cultivator may join a *susu* group of social network members and

through that get access to a lump sum of money to improve upon his/her farming activities. *Susu* is a system whereby a number of people agree to contribute a specific amount of money to a common fund, say at the end of every month. In turn, each member takes the contributed money for their use on a rotating basis.[8]

An analysis of my data reveals that lower class farmers use both borrowing systems more than middle/upper class farmers do. Middle/upper class farmers normally do not need to borrow from others in order to farm, as they have enough resources.

The cultivators in this study had not formed any cooperative organization during my fieldwork.[9] However, it can be inferred from their attitude and answers that if they decide to form cooperative organizations, social networks will play an important role. I asked urban cultivators in Accra, 'If you decided to form a cooperative organization who would be members of the organization?' Their answers are tabulated below.

Table 8.6 Formation of Cooperative Organizations: People to Team Up With

Cooperative Membership	Frequency	%
Network	151	75.5
Anybody	49	24.5
Total	200	100.0

These figures show that 24.5 per cent of the respondents would form organizations with anybody who desired to join. The remaining majority of 75.5 per cent of the respondents preferred to form cooperative organizations with people they know: their current social networks members. The decision to form organizations with members of one's current social network is a rational behaviour since 'when a person chooses from a range

[8] See chapter 9 for a more detailed discussion of credit systems available to farmers.

[9] Earlier attempts at forming cooperative organizations failed under accusations of leadership corruption.

of persons with whom to enter into social exchange transaction he is more likely to choose a man who has rewarded his activities in the past' (Ekeh, 1974:123). The fact that 75.5 per cent of the cultivators would rather form organizations with people already in their social network again indicates that trust is one of the necessary preconditions for the formation and survival of such a voluntary organization. Through the existing social networks the cultivators trust each other. In addition, as earlier mentioned, members of an organization may continue to be members only if they derive some benefits from it. The urban cultivators in this study get some assistance from their social network colleagues even though they do not have any formal organization. On the basis of this, they assume that if they formed a cooperative organization they would gain much more from each other.

Family members or kin are influential in the formation of social networks because in many cases a significant number of the members of a person's social network are kin. In the next section, I will discuss how changes in the family system have affected the social network of the average urban farmer.

8.7 Change in Family Structure and Urban Agriculture

Traditionally, the family system in sub-Saharan Africa is the extended type of family. An extended family is a 'family in which several closely related nuclear families live together as a functioning unit' (Broom, Selznick and Darroch, 1981:575). In this type of family, a kindred of two or more generations co-habit, involving more than one reproductive unit or survivors of such a unit. Also, family members may live 'in neighbouring houses, either within village communities or in more dispersed patterns' (Oppong, 1992:72) but they maintain constant touch with each other. Since it encompasses many kindred, the extended type of family can be termed a macro-level social organization.

An extended family has a great economic value for its members. For example, it may provide farmland for its members, and it is an important source of farm labour. Also, in times of economic or financial crises members receive assistance from the extended family. In addition, it provides child care and other social services for its members (see Obosu-Mensah, 1996). It should, however, be mentioned that since an extended family demands collective obligations it hinders individual independence.

In rural Ghana, the extended family system is still the most common type of system. However, in urban Ghana it has undergone some modifications. These modifications are the result of the fact that most of the time people migrating to urban areas leave their fathers, mothers, uncles, and other family members behind. There are various reasons why family members are left behind during migration but this is beyond the scope of this book. It should be noted though that it is no coincidence that the traditional extended family system has declined in urban Ghana. Modernization theorists and urban developers identify factors like achievement, motivation and a decline in the significance of extended family relationships as crucial for economic development (see Webster, 1991:56). Thus, in a sense, the extended family system is discouraged in urban Ghana. For example, the building plans approved by the government are not conducive for extended family living: traditional building structures with open spaces in the middle are not recommended for urban Ghana.

The modifications in the extended family have resulted in two main types of family systems. The first one can be termed the revised extended family system, and the second type is known as the nuclear family system (or conjugal family system). Both systems have some common characteristics. According to Obosu-Mensah (1996), with the new types of family systems, loyalty to one's extended family or kin is not considered the highest priority. Thus, individual decisions are not necessarily made with reference to the interest of the extended family and individuals are not controlled by the family head. This indicates a shift towards individualism. In addition, with the new family systems it is not necessary for children to follow in their parents footsteps, as far as their profession is concerned.

The revised extended family system lies between the extended type of family and the nuclear family. Typically, it consists of a couple, their children and some other relatives, most often the mother of the wife. It may also include brothers and sisters of the spouses, and people who are not family but close friends, especially people from the home towns or villages of the spouses. The revised extended family system is not typical of urban Ghana but it also exists in some rural localities. For example, the Volta Resettlement Scheme of 1964 transformed the existing extended family system in some rural communities into the revised family system (Obosu-Mensah, 1996). An important difference between this and the extended type is that membership in the former is smaller, while that in the former is larger. In Accra, the revised family system is mostly practiced by open-space cultivators because, as already noted, they are more willing to accommodate new migrants. As shown in Table 8.7, there are more people in the families of open-space cultivators than in the families of enclosed cultivators. For example, 62 per cent of the households of open-space cultivators in this study contain five or more people in the household. Comparatively, only 20 per cent of the households of the enclosed cultivators contain five or more people.

Table 8.7 Number of People in Household of Urban Cultivators

Type of Cultivation	Number of People in Household				
	1-2	3-4	5-6	7+	Total
Open-space	17	40	56	37	150
Enclosed	15	25	10	0	50

The nuclear type of family, which evolved mostly as a result of industrialization and urbanization, encompasses fewer people: husband, wife, and children. It is a micro-level social organization. It predominates among non-farming communities, and obligations between spouses are emphasized. Most enclosed cultivators in this study practice the nuclear type of family.

Many urban residents do not have kin members in the towns and cities where they live. The *loss* of kin is however replaced with friends and other members of their social networks.

As far as urban farming is concerned, the most important feature of the extended family system, which is lacking in both the revised extended family and the nuclear family systems, is the availability of kin labour. Since both these systems are composed mostly of spouses and children, members are not readily able to count on the labour of kin. Thus, the nuclear family system deprives urban cultivators of an extended use of kin labour. However, due to the fact that most urban farms are small, the lack of kin labour does not have much impact on urban agriculture.

8.8 Summary

The discussion so far indicates that social networking is very important in the success of some urban farmers, particularly lower class cultivators because without social networks they would not be able to acquire the necessary capital, land, and other inputs needed for cultivation.

The poorer the urban farmer, the more he/she relies on network members for assistance. However, the assistance a farmer receives depends on his/her network ties. Those with more intense ties receive the most assistance. Data from the field indicate that the intensity of networks of lower class or open-space cultivators is higher than that of middle/upper class cultivators. The network members cultivating close to each other have frequent contact with each other. This frequent contact enhances faster and easier delivery of assistance to needy network members.

Comparatively, the social network of middle/upper class urban farmers contains fewer people than that of lower class farmers. Furthermore, the network of middle/upper class farmers in Accra is looser than that of lower class farmers.

In deciding to relocate to urban areas, rural residents are influenced by the people they already know in the new place. For example, most of the cultivators in this study knew some people in Accra before they moved to the city: they may not have come if they did not have network members in Accra. Information gathered from the field indicates that middle/upper class cultivators are less willing to accommodate their kin who migrate to Accra.

The cultivators in this study had not formed any cooperative organization during the period I studied them. However, they know the people they would form cooperative organizations with if necessary. As expected, in the eventual formation of cooperative organizations they preferred members of their current networks to non-members. This is due to the fact that they are familiar with and trust their current network members.

Eventually, interactions within network groups produce leadership. Individuals who provide the most important needs of the other members of the group become the leaders.

I noted in this chapter that due to changes in the traditional family system in urban Ghana, family members have limited access to the labour of kin. I also noted that this does not affect urban farming adversely, since urban cultivated plots are small in sizes and for that matter do not require extensive labour from kin. Generally, from this chapter one can conclude that social networking and subsequently social exchange may lead to social solidarity.

In cultivating in cities, urban farmers encounter some problems. These problems are discussed in the next chapter.

9 General Problems of Urban Farmers and Ways of Coping

9.1 Introduction

An efficient control of labour, technology and capital is a necessary condition for a successful agricultural production. Thus, an urban farmer's success in cultivation is partly dependent upon the labour, technology, and capital available to him/her. These three factors are interrelated but are not of equal importance.

Capital is primary, and the other factors are dependent upon it. I argue that capital is the most important among the aforementioned factors because a farmer with enough money is able to hire labour and purchase all the necessary technology needed for farming. Generally, urban farmers encounter several problems in their enterprise which is due to the fact that the majority of them do not command adequate resources.

Improved technology increases agricultural production and consequently increases profits and food supply to the population. However, technology is expensive, therefore not every farmer has access to appropriate technology. Appropriate agricultural technology includes the efficient use of farm equipment, the efficient application of fertilizers and chemicals, proper tilling of the land, and the provision of adequate water including irrigation water for cultivation.

Labour too is important. Farming activities, including the efficient use of technology, are undertaken by labour. Labour may be skilled, semi-skilled or unskilled. The more skilled a labourer, the more efficient he/she is. Skill is attained through education and training or experience, and not every farmer is a skilled farmer. An unskilled farmer may hire the services of skilled farmers but due to financial constraints, not every farmer is able to command skilled labour. Many urban farmers in Accra do have problems as far as skilled labour is concerned, however, they do not encounter shortages of unskilled labour. This is because they have access to cheap labour supplied by the numerous unemployed. Moreover, as previously noted some of them use the services of kin, especially newly arrived kinsfolk. Also, on the issue of availability of labour they do not have problems because they cultivate small areas of land. This means they are able to handle their agricultural practices even without the assistance of other people's labour.

In this chapter, I will look at the general problems urban farmers encounter. As expected, these problems are contingent upon the capital available to each farmer. In addition, information provided by cultivators in this study will be used to determine how they are coping with their problems. My main thesis for this chapter is: successful urban agricultural production is dependent upon avoiding and/or controlling a number of problems identified in this chapter: problems associated with land security, water supply, roaming animals and others. Hence, there is a connection between the productivity of urban farmers and the agricultural problems they encounter. A farmer who is able to avoid or solve his/her problems achieves higher productivity. Conversely, a farmer who is not able to avoid or solve agricultural problems experiences decreased productivity. In connection with the above-mentioned thesis, I hypothesize that the more secured a cultivator's land tenure, the more willing he/she is to invest in his/her farming activities. All things being equal, the more a cultivator invests in his/her cultivation, the more he/she reaps.

In Accra, I asked the respondents to mention their problems according to importance. Table 9.1 summarizes the responses.

It should be noted that 41 cultivators mentioned they do not encounter any problems related to cultivation. It is pertinent to note that all of these cultivators are middle/upper class or enclosed cultivators.

Table 9.1 Main Problems Identified by Cultivators

Problem	Frequency	%
Lack of Resources	57	28.5
Land Tenureship	52	26.0
No Problem	41	20.5
Pilfering	18	9.0
Stray Animals	16	8.0
Gutter Water	15	7.5
Pests	1	0.5
Total	200	100.0

9.2 Lack of Resources

The cultivators involved in this study mentioned lack of adequate resources as their most important problem. Under resources they mentioned logistical problems like lack of capital, high cost of input, lack of credits, technological know-how, and lack of a reliable market. Most of the cultivators mentioned that since they didn't have other reliable sources of income, they don't have enough capital to invest in their farming activities. They noted that without enough money they were not able to acquire the appropriate tools, chemicals and fertilizers needed to improve upon their farming activities. On the problem of lack of capital, the cultivators call on the government to institute a financial organization that caters specifically for the financial needs of urban farmers. They suggested alternatively that urban farmers should be given the same recognition as rural farmers by the Agricultural Development Bank, and extend soft loans to them.

Most of the open-space cultivators complained that the prices of agricultural inputs like chemicals, fertilizers, seeds and tools were too high. The least affordable input was chemicals, followed by seeds and fertilizers. The cultivators explained that the prices of these inputs were high since they were imported. One open-space cultivator mentioned that,

> Since private people [businesses] got involved in the importation and sale of agricultural inputs prices have been going up unabated. You see, the government is gradually moving out of this business leaving it in the hands of private interests. These people want to make maximum profits so they keep on increasing the prices as they want. There is no more government subsidy, and one thing is that the prices of agricultural inputs fluctuate according to the American dollar rate at the forex bureaux.

Some urban cultivators suggested that the government should reintroduce the subsidization of agricultural inputs. In addition, Ghana should pay as much attention to the manufacture of agricultural products or equipment as it gives to the manufacture of industrial products.

On the high cost of fertilizers, I suggested to the cultivators to use manure or compost since they may get this free of charge. They admitted that compost, is as effective as fertilizers but they noted that it is heavy to move from the main source at the Accra Metropolitan Assembly's composting site to their farms. When I suggested that they could make their own compost from their refuse or garbage, a few questioned the safety of compost. As one of them put it,

> I cannot carry garbage all the way from my house to my farm. Secondly, it is not hygienic to handle garbage in that way. Garbage is meant to be discarded as soon as possible.

Most urban farmers do not have enough technical know-how in urban agriculture to improve upon their yields. In chapter 6, I established that most of the cultivators come from rural background, meaning they have experience in rural agriculture. While a rural background may be necessary to induce one into urban farming, rural agricultural practices are not necessarily applicable to urban agriculture. Unlike rural agriculture, urban agriculture is more intensive in that a farmer relies on more elaborate methods to realize good yields from a smaller area of land.

Moreover, the crops mostly grown by urban cultivators are different from those grown by rural cultivators, thus necessitating different methods of cultivation. For example, a rural cultivator who cultivates cacao plants clears his farm of weeds occasionally but does not water the crops. An urban cultivator who grows salad, for example, has to water the crops till they mature. In addition, he has to observe strict distances between his crops in order to attain maximum yield. It is also necessary for a vegetable cultivator to apply fertilizer or manure to his or her soil in appropriate quantities, while a rural farmer may not apply fertilizer to his soil when cultivating, for example, cacao plants.

It is a matter of observation that as people increase their input to a unit area of land, their output also increases. It is also known that after a time, increasing input does not necessarily increase output proportionally. For example, 'doubling fertilizer application will not necessarily double output' (Grigg, 1995:74). Like the economic law of diminishing returns, an increasing application of, say, fertilizer, will increase yield only till a certain point and then the comparative yield would drop. Actually, in agriculture an increasing input may be detrimental to growing crops and animals. If a farmer continually applies more water to his/her crops there would come a time when there would be too much water, subsequently causing waterlogging and wilting. Hence, urban farmers need adequate information on good agricultural methods.

Vital agricultural information which urban farmers lack should be provided by specialists. The lack of technological know-how emphasises the need for contact between urban farmers and Agricultural Extension Officers. However, as mentioned in chapter 7, lower class cultivators, those who need the services of Extension Officers most, do not receive it.

Another problem mentioned by the cultivators in Accra as a logistical problem is marketing. Generally, the cultivators said they do not have problems selling their products. They sell in small quantities to consumers but they prefer to sell in bulk. However, there is no organization that buys in bulk from them. The main problem they identified with selling in small quantities is that since the money comes in small amounts, they are not able to save. In Ghana, there is a Ghana Food Marketing Board which buys from rural farmers in bulk and retail to consumers in towns and villages. Some urban cultivators in this study suggested that the Board should buy from them in bulk.

At the time of this study, a few individuals have opened vegetable shops around Accra, and they bought in bulk from some urban farmers. The trend seems to be an increasing number of such shops as the years go by. Consequently, I believe that in a few more years many farmers will have the opportunity to sell in bulk.

Lack of credit is another logistical problem mentioned by the cultivators in this study. A cultivator who has not got his/her own capital to spend on his/her farm would have to borrow. It is a matter of commonsense to mention that in the event of lack of own capital, the farmer who is able to secure credit is more likely to improve upon his/her yield than one who is not able to secure credit.

Many urban cultivators in Accra don't have enough money to improve upon their farming (see section 9.3). Therefore, it is important for them to borrow money from elsewhere. To find out whether the cultivators ever thought of borrowing money from banks I asked open-space cultivators, 'Do you need some money to improve upon your farming activities?' Further, I asked 'Where will you get the money from?' See Table 9.2 for the answers given to the former question.

Table 9.2 Number of Open-space Cultivators Who Need Credit

Credit Needed?	Frequency	%
Yes	133	89
No	30	11
Total	150	100

Eighty-nine per cent of the respondents mentioned that they need credit to improve upon their farming activities. However, only one of them mentioned the bank as the institution he will contact for a loan or credit. The urban cultivators in Accra showed that they know a lot about banking institutions in Ghana. Therefore, one of the questions I asked those who mentioned lack of capital as a hindrance to improving their yields was, 'Why don't you borrow some money from a bank to improve upon your cultivation?' As expected they mentioned that they could not pledge their farms as security for bank loans. As one of them put it,

> The lands we farm are not ours so we cannot pledge them for loans. No bank will approve our loan applications if we tried. Apart from that, I don't think that bank officials see urban cultivation as a lucrative venture. It is funny because they (bank officials) buy vegetables from us everyday. And some of them are themselves involved in urban cultivation.

As already mentioned, many urban farmers want to improve upon their yields but they are not able to do so due to the lack of capital. Therefore, they should borrow money from elsewhere. In Ghana, there is an Agricultural Development Bank (ADB) which caters for the banking needs of farmers. ADB extends a specially lenient loan conditions to farmers, so this is the best bank urban farmers could contact for loans. The problem is that officially, urban agriculture is not considered 'normal' farming because, among other factors, cultivators may lose their farmland any time. Consequently, urban farmers do not qualify for the special conditions of borrowing extended to rural farmers.

In their efforts to increase their capital to improve cultivation, some open-space cultivators in Accra secure credit from non-institutional sources including family members, members of their social networks and private money-lenders, sometimes, at exorbitant interest rates. The ADB (1975) identifies three main charges on loans provided by private money-lenders. These are commitment fees, interest charges (rates), and debt servicing fees. The payment of commitment fee which varies from area to area, is a sort of 'thank you for the loan.' Thus, this is not counted towards loan repayment. According to the ADB, 'interest rates vary between 33½ and 100 per cent' (ADB, 1975:37). The more urgently a person needs a loan the more he/she pays as interest. Thus, 'the 100 per cent interest rate (called 'Nsianimu' in Twi) is usually charged when the lender feels that the borrower is really in need of money' (ADB, 1975:37). The 'debt servicing fee [which is a percentage of the amount owed] is levied only in the situation where a loan is not fully amortized at the end of the pay-back period and repayment has to continue thereafter' (ADB, 1975:37). Like the commitment and interest fees, the debt servicing fee, which is scaled down in proportion to the diminishing balance of the outstanding loan, does not affect the principal amount owed in the sense that payment of debt servicing fees does not decrease the amount owed.

Normally, private money-lenders lend money to people they know. If a money-lender has no idea (e.g. credit worthiness) about the prospective borrower, he/she 'ascertain[s] the credit-worthiness of the applicant' (ADB, 1975:34). Residing in the community in which he operates, a money-lender is able to form an idea of the repayment capability of the borrower. Due to this, and to the fact that they are in close association with their debtors, 'money-lenders are more successful in their business than the other non-institutional and institutional credit agencies' (ADB, 1975:29).

It should be noted that non-institutional money-lending is at times surrounded by unclear terms or conditions. Therefore, they are the source of some conflicts. In the Agricultural Development Bank's report, a money-lender is said to be regarded by a section of the society as 'the inconsiderate friend of the farmer who comes to his aid in times of need only to squeeze him dry in times of plenty' (ADB, 1975:4). However, another section of the society considers the money-lender as 'an asset to the village farming community and he is held in high regard. Besides, he is approachable and he is ready to lend at short notice' (ADB, 1975:29).[1] According to the cultivators in this study, borrowing from money-lenders is not adequate and not preferred.

Another source available to urban cultivators for raising capital is called *awowa*. It is a sort of an arrangement in which a farmer pledges his/her farm in order to borrow money from a private lender. Traditionally, it involves perennial cash-crops like cocoa. Whilst the farmer is repaying the loan, the lender maintains the pledged farm. The lender harvests the crops growing on the farm but they are not counted toward repayment of the loan. Consequently, until the money is physically repaid, the lender enjoys the benefit of 'owning' the farm. The pledged farm is therefore just an interest on the loan.

Over the years, *awowa* has taken different forms to suit also annual crops, which are considered to have less value. When it involves annual crops, the loan may be repaid in cash or in kind. When the loan is to be repaid in kind, a cultivator pledges his/her farm to a lender on the agreement that the lender may harvest the crops when they mature. There are two types of this system. Under the first one, the growing crops become the outright property of the lender. He takes care of the crops. Therefore, when the crops mature he harvests them to pay for the loan. In this case, a price is put on the crops during the time the loan is negotiated, that means before the crops mature for harvesting. With the second type, the cultivator (the borrower in this case) entrusts the harvesting of his crops to the lender who sells them after harvesting, and the amount lent out plus the interest is deducted. If there is any money left after the deductions it is given to the cultivator.

The main difference between the two types of *awowa* is that with the former, the growing crops are entrusted to the lender. The lender takes care of the farm. This is a sort of outright sale of the growing crops to the lender. However, with the latter type the cultivator takes care of the

[1] These contradictory remarks about money-lenders indicate how controversial they are: they are essential to the survival of some, and 'hated' by others.

growing crops until maturity, when the lender comes in to harvest.

Awowa is hardly used in urban Ghana, though it is common in rural Ghana. Only three of the respondents in this study borrowed money under the *awowa* system. They were compelled into this agreement by deaths in their families. According to them, they spent a lot of money on the funeral and other arrangements, so they had to borrow money at a time when their crops were not yet matured.

While *awowa* system of money borrowing may provide a cultivator with the needed money to invest on his/her farm, it is also the source of some conflicts. The conflict is mostly over the value of unharvested crops. As economic beings, both the lender and the borrower want to make maximum profits, so each tries to get the best out of the bargain. Consequently, when a cultivator insists that his/her crops are worth, let's say C100,000 the lender may say they are worth C70,000, thus causing friction between the parties. The ADB notes that, 'the creditors' prices are usually lower than the ruling market prices' (ADB, 1975:41). Conflicts may also arise when pledged crops fail. On concluding the deal, crop failure is not normally taken into consideration. If the pledged crops fail, who is to take the burden, the cultivator or the lender? When pledged crops fail, the cultivator may say the crops had been entrusted to the lender, so it is the latter who should bear the loss. The lender, on the other hand, may say that he/she was promised the harvests, not the failure, so the cultivator would have to bear the loss.

The *awowa* arrangement for money borrowing may be made between:

a) an illiterate farmer and an illiterate lender;
b) an illiterate farmer and a literate lender;
c) a literate farmer and an illiterate lender; and
d) a literate farmer and a literate lender.

The most reliable *awowa* arrangement is ensured if both the lender and the cultivator are literate. In this case they may write the entire contract or agreement down.

In a conflict over crop failures and over the prices of growing crops the money-lender usually wins especially if he/she is literate. A money lender is richer than a borrower, and wealth is a source of power: he/she 'wield[s] a lot of influence' (ADB, 1975:33). A borrower is more dependent on a lender than vice versa. In addition, there are more borrowers than lenders, so if they encounter problems with a particular borrower, there are many other borrowers they can deal with. Furthermore, if a borrower is blacklisted by a lender, other lenders desist from dealing with him/her.

Conflicts that might ensue are attributable to the fact that most *awowa* transactions are not written down. The only way to prevent conflicts is to write down the terms of the transactions. This would be easier if both the lender and the borrower were literates.

The only advantage of *awowa* is that it makes money available to a cultivator at a time of need. However, it should be noted that like the example above, where money was borrowed for funeral expenses, normally the money is borrowed for use on non-farm activities rather than to improve upon cultivation.

Another type of credit available to urban cultivators in Accra is called *susu*. It is an informal association of voluntary contributors of money into a common fund. Depending on the agreement, the financial contribution is either on daily, weekly or monthly basis. Each contributor draws from the fund at a previously agreed time. Brown (1993) describes *susu* as a 'rotating savings group, in which each member deposits a small amount of money regularly into a central fund, the whole of which is given to each member in turn' (Brown, 1993:3). For example, ten people may be involved in a *susu*. They may agree to contribute C1,000 at the end of every month. At the end of every month one of them, in turn, takes all the contributions. Thus, instead of C1,000 in hand at a particular month, a *susu* member in our example may have C10,000 at that particular month. *Susu* which is widespread in both rural and urban areas of Ghana, may have originated from 'the [traditional] granting of loans by societies to members as a method of relieving existing indebtedness...' (Anyane, 1963:186).

Susu is very beneficial when one needs a lump sum of money. Another advantage is its 'non-adherence to formal security nor specific office opening hours' (Brown, 1993:4). In addition, there is 'a higher probability of obtaining an informal loan [through the susu system]' (Aryeetey, 1992:14). However, it has some disadvantages. In the first place, contributors do not earn interest on their money. Second, 'inflation causes cash savings to fall in value' (Brown, 1993:3). Thus, by the time a contributor collects the lump sum of money its value might have fallen. It also requires a dedicated person to go around to collect members' contributions. This means, each *susu* group is normally limited to a few members who live in close proximity.

Many urban cultivators in Accra join *susu* groups. When I asked, 'Are you a member of any *susu* group?' I got the response shown in Table 9.3.

Table 9.3 Number of Cultivators Who are Members of *Susu* Groups

Type of Cultivation	*Susu* Membership Yes	No	Total
Open-space	90	60	150
Enclosed	6	44	50
Column Total	96	104	200

It is clear from the above table that comparatively *susu* is more attractive to open-space cultivators than to enclosed cultivators. Of the 150 open-space cultivators in the Accra study, 90 (60 per cent) belong to *susu* groups. Within enclosed cultivators, only 12 per cent are members. This implies that while middle/upper class cultivators need not join any groups to raise money for their farming and other activities, lower class cultivators have to. The fact that more open-space cultivators are involved in *susu* is an indication that they are economically poorer than enclosed cultivators.

Finally, some cultivators in this study mentioned that they borrowed money from members of their network groups and kin without paying any interest. The amount involved is normally not huge, and it is generally for a shorter duration.

Though the non-institutional credit systems urban cultivators use are at times the source of conflicts, they bind cultivators as well as others together in the form of cordial relationships.[2] Generally, compared to formal sources of credit, informal sources of credit 'are [more] convenient, available locally, require no documentation and can provide credit quickly' (Upton, 1996:160).

9.3 Problems of Land Tenureship

An indication from Table 9.1 is that the respondents mentioned land as their second most important problem, after lack of resources. I should mention

---

[2] The informal credit arrangement may be between two farmers or between a farmer and a non-farmer.

that lack of resources shows up as the most important problem because it incorporates many different variables, including credit and logistical problems like capital and technical know-how. Thus, land would have been the most important problem, had it not been for the fact that lack of resources incorporates many variables.

When the cultivators were asked to explain how land posed problems for them, they mentioned lack of access to land as the foremost problem. This was followed by the problem of land security. For the cultivators, the problem of land tenureship is the result of inefficient or lack of government policies on urban agriculture. They explained that a cultivator may cultivate an idle land without any problems, provided the owner is not ready to develop it. However, as soon as an owner decides to develop his or her land, cultivators are ejected from it without proper notice. In some instances, crops growing on the field are destroyed to make way for construction.

Traditionally, many Ghanaians have access to farmland in rural Ghana. However, it is difficult, if possible at all, to move land for agricultural purposes. As noted elsewhere, 'labour entrepreneurial skills and inputs such as fertilizer or machines can all be moved from one place to another; land cannot' (Grigg, 1995:69). This implies that a person with large areas of land at one place may lack land at another place. Consequently, many of those who possess tracts of land in rural areas may lack land in urban areas. During my fieldwork one urban cultivator reiterated this by saying,

> This world is unfair. I have large areas of land in my village but look at me. Here in Accra I don't own even a hectare of land. I wish I could transport my land to the city.

The issue of land tenureship in many African countries is still very volatile. Much of the land is communally held, so plots are not sold to outsiders. In Ghana, land is never sold but leased out normally for a period of ninety-nine years. After this period the lease should be renegotiated. Legally, the original owners or their descendants can decide not to renew the lease.

Owing to the high prices of land in Ghanaian towns/cities only middle/upper class people are able to purchase land.[3] Consequently, as

[3] The most appropriate word should be lease (not purchase).

already noted, many urban cultivators don't own the land they cultivate. It should also be re-emphasized that land may not be bought in urban Ghana for the purpose of farming, so this is another constraint to urban cultivation.

Middle/upper class urban residents, by virtue of their capability to purchase land, cultivate on land they could call their own (see chapter 7). However, since lower class urban residents cannot afford to purchase land, they cultivate on pieces of land that do not belong to them. This indicates that while middle/upper class urban residents have security on the land they cultivate, lower class urban residents do not have any security on the land they cultivate. While all the middle/upper class cultivators in the Accra study owned or rented the land they cultivated, only three of the lower class cultivators (here, open-space cultivators) owned the land they cultivated. Therefore, land security is a serious problem for lower class urban cultivators. They cultivate public and private land whose owners they might not know. Similar observations have been made in other sub-Saharan African cities. For example, in Nairobi Freeman (1991) observed that about half of the urban cultivators were using public land or did not know who owned the land they farmed. In a study in Kampala it was noted that 'fewer than 10 per cent of the respondents in the survey own the land on which they are farming' (Maxwell and Zziwa, 1992:37).

My data show that among lower class cultivators the educated are more likely to know the owners of the land they cultivate. Consequently, they are more likely to make arrangements with land owners for the use of land. Of the ten open-space cultivators who said they have made some arrangements with the owners of the land they cultivate, seven have secondary or tertiary school background. The remaining three have elementary educational background.

The fact that many farmers do not own the land they cultivate has been detrimental in one important way. This is made clear by Tinker (1994) when she writes that 'the uncertainty created by expected harassment keeps farmers from investing in soil and crop improvement' (Tinker, 1994:xiv). For those cultivating on other people's land there is always the threat of eviction so they do not make long-term investments. The fear of eviction, was mentioned by many open-space cultivators in Accra. I asked open-space cultivators how much they invest on their farms. However, the answers obtained cannot be used in this study, since they tend to be unreliable. This is due to the fact that the respondents do not keep records of their farming activities they do not know how much they invest on their farms. The only conclusion to draw from their answers is that those who do not have any secured land tenureship invest less on their farms. One explanation is that the plots of land they cultivate may be taken from them at any time so, there was no point investing their money in them. People in this category noted that if they invested more in their farms they would increase their productivity. A cultivator at Abosey Okai made this clear when he said,

> A friend of mine has a tenureship arrangement with his landlord, so there is no fear of this man being evicted from the land without prior notice. Consequently, he invests more money into his farm. I don't even know the owner of the land I farm, and he or she may reclaim the land any time so I dare not pump my money into the farm. You see, despite the fact that I'm more hardworking than my friend, he makes more money than I do. The determining factor is, he can invest on his farm and I cannot.

During the period of my fieldwork, I witnessed two groups of cultivators lose their farms to the bonafide owners of the land. The way these occurred make it important for them to be recounted here. The first one was at Tema, and it happened on Sunday, January 15, 1995. According to the cultivators,

> When we were here (in the farm) yesterday, a group of four strangers came to the land. They did not even greet us. They walked around the area while discussing an issue. They pointed fingers at certain points of the land. Now we know that they were pointing to the boundaries of their land. They left without saying anything to us.

The farmers left for their homes later in the evening, wondering whether the four men were the owners of the land. Anyway, there was one thing they were sure of: even if the people owned the land, they were not thinking of developing it now. If they were, the cultivators thought, they would have mentioned this and asked them to vacate the area. However the following evening, the children of some of the cultivators ran to them saying,

> Our crops were being levelled down by a bulldozer. When we rushed to the farm we could not believe what we saw.

> How can some people be so wicked? Our crops were at a maturity stage, and they could have given us some time to harvest them. I don't understand why they could not tell us this yesterday. When we asked them why they were destroying our crops they told us that the land belongs to them so they had the right to clear it of 'weeds' any time.

It was a pathetic scene to see the cultivators trying to harvest whatever they could before the bulldozer destroyed everything. It was a lost battle since the machine was faster than them. The land owners had rented the bulldozer on an hourly basis, so the faster they worked, the cheaper it was for them. The area was cleared for the construction of a hotel.

The second incident, which is similar to the first one, happened on Thursday, May 11, 1995, in Accra opposite the main entrance to the University of Ghana. Here, the cultivators came to their farms in the morning just to see their crops being destroyed. This land was cleared for the construction of a petrol (gas) station.

Here is another cultivator's ordeal which happened in May 1992. He narrated,

> I was told the land was meant for road construction. Some officials came and took measurements of the area. At this time I had planted rice which was almost matured for harvesting. I also had plantain, cassava and sweet potatoes growing on the land. The officials [of Construction Pioneers] took pictures of my farm and left for their office. I followed up to their office, and asked them what compensation I will receive from them. They told me they will come and assess the cost but they never turned up. Without any prior information, they came and cleared all my crops. I was later asked to come to their office for compensation but up till now (1996) nothing has been given to me. Any time I go to their office I'm asked to come another day. I'm now fed up and thinking of forgetting about the whole issue.

The fact that land insecurity is a major problem for urban cultivators can be made even clearer by a study in Nairobi, which concluded that most of urban residents who quit urban cultivation did so due to eviction or otherwise losing land-use rights (see Mazingira, 1987:280). In Accra, the cultivators mentioned eviction from land as the second most important factor that may compel them to quit urban cultivation. Most of the non-cultivators, including some officials said, they were not cultivating because of unavailability of farmland.

The land ownership problem is very discouraging for many cultivators because they rely to a large extent on their farms. For most, urban cultivation is a permanent issue, so they would rather have a secured land for cultivation.

I asked urban cultivators how the land problem could be solved. They suggested that urban planners should take urban cultivation into consideration when planning urban areas. I prompted them to speak more on this issue by asking them, 'In the light of increased urban population, how do we solve accommodation problems if areas of land are set aside for cultivation instead of constructing accommodations?' In summary, they said the land issue could easily be solved by expanding the perimeters of Ghanaian towns and cities. By that they mean peri-urban areas should be incorporated into towns and cities, so that one may not talk of shortage of building areas. They mentioned that with the increasingly efficient transportation systems it is not necessary for all urban residents to stay close to city centers. Actually, the current trend in Accra is that many people are building in areas which are not really part of Accra. For example, many people are building houses and moving to villages like Amasaman and Kasoa, which are at the fringes but outside the boundaries of the city.

The above suggestion is very rational when one considers that competition over land is tenser when there is land scarcity. By expanding the perimeters of cities, many tracts of land would be made available for development, thereby lessening the competition for available land. In Ghana, land is not in short supply. Consequently, city officials should be able to expand the perimeters of towns and cities. They should be able to include urban cultivation in the plans of cities.

9.4 Pilfering

Stealing of crops is an important problem, the cultivators in Accra mentioned. According to them, they engage in cultivation in order to produce food for their own consumption, and mostly for sale.

Consequently, their motivation to cultivate decreases when their crops are stolen. According to the cultivators, lower motivation has led to lower productivity in some instances, thus lowering their profit margins. It follows, therefore, that a cultivator who has little or no pilfering related problems gains more from urban cultivation than one whose crops are target of thieves. Cultivators in almost all the areas of my study mentioned stealing as a problem. Most of the cultivators mentioned they are helpless as far as stealing is concerned. As mentioned in chapter 7, since their farms are along roads, passers-by easily steal crops without being detected.

The cultivators at Dzorwulu expressed disappointment with the way the police handled thieves arrested in their farms. This is what one of them said,

> It is very disappointing when the police refuse to prosecute thieves we apprehend. They don't prosecute them for two reasons. In the first place, the police don't care about us, they don't see us as important people. For them, we are just poor people who do not matter. In the second place, some police officers are so corrupt that they take bribes from the thieves and let them go.

To solve the pilfering problem, some of the cultivators, especially those at Dzorwulu, have decided to, form vigilante groups to keep watch over their crops from maturity until they are harvested.

9.5 Stray Animals

Stray animals, especially ruminants that destroy growing crops, are also a matter of concern for urban cultivators in Accra. Like any of the other problems urban cultivators encounter, this reduces their motivation and yields. This is what a cultivator said,

> It is very disappointing to go to your farm and find out that goats have grazed on your crops, and/or pigs have destroyed your beds. You get so discouraged that you think of stopping cultivation but, well, we should not stop cultivating. If anything at all it is the animals which should be stopped from destroying our crops.

And as with the pilfering problem, the cultivators feel the authorities are not doing enough to prevent animals from destroying crops. In chapter 5, I noted that by law it is not allowed for animals to be left to roam the streets of Accra. The same by-law stipulates that animals left to go astray would be confiscated. However, as the cultivators acknowledged, officials do not enforce this by-law, so roaming animals are not confiscated. Some of the cultivators suggested that officials of Accra Metropolitan Assembly should catch roaming animals and impose heavy fines on their owners. On their own, some cultivators at Abossey Okai remain in their farms till late in the evening when roaming animals have returned to their pens.

9.6 Gutter Water

The majority of urban cultivators grow vegetables. Vegetables grow fast and require a large quantity of water to grow properly. Therefore, it follows that urban cultivators need adequate sources of water supply. In Accra, the main supply of water to the general public is tap (piped) water. However, tap water is not a convenient source of water for urban farmers because it is expensive and unreliable.

Only 15 cultivators mentioned inadequate supply of water as a problem. Consequently, for the average cultivator in Accra water is not a major problem because they use gutter water which is available all year round. The issue, as raised in chapter 5, is whether gutter water is a safe source of agricultural water.

The problem is that the water is untreated, and consequently, can be the source of disease outbreak, thus confirming the fears of some Ghanaian officials. A suggestion of this (i.e. disease outbreak as a result of the use of gutter water) was vehemently protested by some cultivators. At Cantonments one of them said,

> We have been using this water for cultivation over many years and no customer has ever complained of becoming sick after consuming our crops. In addition, we farm near a health clinic, and if the water we are using was bad the physicians would have told us this. They see us use gutter water everyday but they have never shown any negative concern about it.

It may be true that no consumer has ever complained or objected to the use of gutter water by the farmers but there is some concern among some residents of Accra about its use. When non-cultivators were asked about their main concern about urban agriculture, some of them mentioned the use of gutter water. A lady who resides in Darkoman, a suburb of Accra said,

> Whenever you have the time I will take you to an area where a man is cultivating, and you will see for yourself the type of water he uses. Anybody who sees the water he uses will not touch his crops. No wonder his wife sells the crops in Accra central, far away from the cultivating area. I don't think the man himself consumes his crops.

Cultivators in Accra and other cities throughout the developing world should take notice of the outbreak of cholera in Santiago as a result of the use of untreated water on farms (see chapter 5). In order to avoid disease outbreak, government officials should occasionally inspect the water used by urban farmers. I should mention that some cultivators have sunk wells at their places of cultivation. It should, however, be noted that the sinking of wells is very limited because they incur the anger of land owners.

9.7 Pests

Only one cultivator in this study mentioned pests as a major problem. It is not clear whether the low incidence of pests in Accra is a result of good agricultural practice or the absence of vegetable pests in general. Some of the cultivators explained that there had been minor incidences of pest attacks on their crops but they were controlled with chemicals bought from Agro-Chemical shops of the Ministry of Agriculture.

Every cultivator's aim is to produce as much as possible and to attain this aim, problems as discussed above should be eliminated or minimized. On a general note, I asked the cultivators the type of assistance they needed from the government in solving their problems. From their answers I identified two main views, as to how and where the government could assist. One view is that there is little or no problem of food production. In this view, the problem is with distribution and marketing. Consequently, any help from the government should be at the level of distribution and marketing. This notion is deduced from cultivators' answers like, 'buy our products in bulk' and 'offer better prices for our crops' to the question of, 'How can the government improve upon urban agriculture?'

Previous attempts by the government to intervene in food distribution did not succeed and the structures that brought about the failure are still in place (see chapter 4). Consequently, it does not seem appropriate for the government to get involved in distributing the food produced by urban farmers. Moreover, increasingly Ghana is privatizing its ventures so, it does not seem feasible that the government will get entangled in food distribution.[4]

The other view, which is stronger than the previous one, is that the most important problem encountered by urban farmers is poor or low production. This was deduced from answers like 'lack of and/or high costs of chemicals', 'problems with land tenureship,' 'lack of improved seeds,' and 'low yields.' In line with this view is the suggestion that the government should assist in increasing urban agricultural yields. However, the cultivators and government officials noted that Ghana's previous involvement in food production through now defunct institutions like the State Farms Corporation, the Workers' Brigade, the United Ghana Farmers' Co-operative Council and the Young Farmers' League did not work (see also Hansen, 1989). Consequently, they do not expect the government to get involved directly in food production. Therefore, they suggested one other main way through which the government could intervene to increase urban food production.

The suggestion put forward by the cultivators is that through laws, the government should make unused land available to cultivators. Furthermore, the government should subsidize agricultural inputs and extend credits to urban farmers. Thus, the government should create more favourable conditions for urban farmers.

9.8 Summary

In this chapter, I identified the problems encountered by urban cultivators. These problems are insecure land tenureship, lack of land and other resources, pilfering, use of untreated gutter water, the destruction of crops by roaming animals, and on a less significant level the problem of crop

---

[4] The State's Ghana Food Distribution Corporation is being privatized.

pests. Urban cultivators use different methods to solve their problems. For example, to make money available for investment on their farms, some of them join or form *susu* groups. To improve their yields through improved methods of cultivation, some urban cultivators take the initiative to seek assistance from Agricultural Extension Officers. And to protect their crops from thieves, some of them keep watch over their farms when the crops mature.

It was noted in this chapter that cultivators who are able to control the above-mentioned problems increase their yields. Consequently, it is the goal of all cultivators to minimize or control problems that may decrease their yields.

In the next chapter, I will draw a short conclusion and recommend some policies that may improve upon urban agriculture.

10 Conclusion and Policy Recommendations

10.1 Introduction

One of the main assertions in this work is that urban agriculture is increasing in sub-Saharan Africa. As long as rural residents with a farming background migrate to urban areas, as long as the urban formal sector is unable to employ all the migrants to urban areas, and as long as many urban residents do not earn enough income from the formal sector to cater for themselves, there will be urban agriculture. Moreover, as urban agriculture becomes more profitable, some middle/upper class urban residents will farm both for sale and for their own use. When they cultivate for sale, middle/upper class urban residents will buy and exploit the labour of small-scale cultivators.

As already mentioned, urban agriculture is a permanent feature in Ghana and other sub-Saharan African countries (see chapter 2). It is therefore necessary for proper measures to be taken to support and improve upon it. Consequently, one of the most important issues at stake is about how to make the practice viable. In line with this position, I will give some recommendations. However, before that I will conceptualize the possible future trend of urban agriculture.

10.2 Future Trend of Urban Agriculture

In order to be able to suggest some recommendations, I have to look at the form of urban agriculture. Several researchers and institutions have concluded that the practice of urban agriculture is increasing and will continue to increase (UNDP, 1996; Mougeot, 1994; Maxwell and Zziwa, 1992; Freeman, 1991). In this work I have indicated that I agree with the projections of these researchers. It is, however, not enough to simply make this prediction. Researchers should examine if the practice will maintain its current form or whether it will take a different form in the future. The factors that would bring about change in the current form of cultivation should also be ascertained.

Colonial and past Ghanaian city planners did not take urban agriculture into consideration. Unfortunately, though present planners condone urban agriculture, they follow the same tradition. During my fieldwork I asked the officials involved in this study whether current planners take urban agriculture into consideration when planning and expanding Ghanaian cities. All of them said no, and they did not see that as a reality in the future. An official of PPMED mentioned that,

> Urban agriculture is becoming more important, and many people are getting involved in it. However, I do not think it will be incorporated into city planning. I cannot imagine any planner setting aside some areas in a city for agricultural purposes. You see, urban lands are in high demand for other purposes. They are too expensive to be used specifically for agriculture. I cannot imagine any Ghanaian buying land in Accra, for example, for cultivation. Many people still think that agriculture is best done in rural areas.

In 1996, I visited new public and private housing projects like Sakumono Estates and Adenta Estates. I realized that urban agriculture was not taken into consideration when these estates were planned so there was little agriculture at these places.

It is obvious from this work that urban agriculture as practiced in Ghana and many other sub-Saharan African countries is basically on a small-scale. It is mostly practiced by urban residents who do not own or have a secured tenureship over the land they cultivate. Furthermore, the Ghanaian government is not directly involved in urban agriculture and does not directly make allocation of resources available to urban cultivators. Thus, urban agriculture is not included in the programmes of the government.

Practiced on a small-scale, many urban residents have the opportunity to engage in urban agriculture. And the absence of government regulation gives the farmers the freedom to farm the way they deem appropriate. However, since urban agriculture is practiced on a small-scale, the government is not able to generate any taxes from it. Moreover, since it is not regulated, good agricultural practices are not always adhered to. For example, some cultivators still use the hazardous Gammalin 20 (DDT) on their crops. Also, due to lack of government regulation urban cultivators are not protected from land owners who claim their lands whenever it suits them, without taking growing crops into consideration. Thus, small-scale urban agriculture has both its pros and cons.

On the basis of findings in the field, I argue that the practice will assume a different form in the future. One of the most important factors that will change the nature of urban agriculture as it is practiced today is an increased development of urban Ghana. Urban agriculture has to compete with other uses of urban land. In this competition, by design, urban agriculture is placed at a disadvantage. For example, while other developers like institutional and residential developers can buy plots of land for their purposes, urban farmers are not permitted to. One cannot buy a plot of land in Accra purposely for agriculture. The fact that urban residents are denied from purchasing land for farming affects the future trend of urban agriculture. Another factor is the commercialization of the practice. Higher vegetable prices would stimulate urban agricultural development in that some middle/upper class urban residents would be attracted by the profits to be made from urban agriculture. Therefore, some capitalists would appropriate the innovation of small-scale farmers and proletarians, meaning there would be a change in the relations of production (see Section 10.4).

There are two major possible trends, depending on whether the government makes a conscious intervention or not. For lack of better concepts I have termed these options Government Regulated Urban Agriculture (GRUA), and Private Commercialized Urban Agriculture (PRICUA).

10.3 Government Regulated Urban Agriculture

With this alternative I assume that, although the officials involved in this study say otherwise, the government will intervene and regulate urban agriculture, particularly in the area of land tenure. I further assume that, like present officials, future government officials will not want urban agriculture to disappear. In chapter 5, I established that Ghanaian government officials appreciate the importance of urban agriculture and condone it.

Due to the importance of urban agriculture to the economy, and if present and future government officials continue to condone the practice, I expect the government to put in place 'zoning and building regulations [that] could be used to shift urban development planning towards productive landscape, thus opening up opportunities for urban agriculture' (UNDP, 1996:246). There are a variety of methods available to the government to make land readily available to urban cultivators. I draw upon Sanyal's (1986) suggestions. According to him the government may,

1. give farmers access to vacant public land;
2. induce owners of private land to allow temporary access for farming;
3. put land around public facilities, such as schools, ports and hospitals to farming use;
4. improve land for agriculture and aquaculture by dredging, filling, levelling, terracing and so on; and
5. design site/service areas for squatters and other low-income residents to provide them with farming space.

Over the years, the government may provide a guideline for lease agreements in order to ensure land tenure security. To be successful, the lease document should specify the duration of lease and other necessary agreements.

Apart from providing a lease guideline, the government may demarcate certain areas in Ghanaian towns/cities for farming. The government would then have to allocate these areas to current and prospective urban farmers. To efficiently handle this, urban agriculture should be put under an institution like the Ministry of Agriculture or Town/City Assemblies. As mentioned in chapter 5, currently urban agriculture does not fall under the jurisdiction of any institution. UNDP notes, 'in most cities, provinces and countries, urban agriculture does not come under the exclusive agenda of any ministry or government department; it therefore falls between the cracks' (UNDP, 1966:239).

Private land owners will be more willing to allow others to cultivate their lands if they are given some incentives. Consequently, apart from the above-mentioned ways of government intervention in urban agriculture, the government may give land property tax relief to owners of private lands who release them to others for urban farming.

With most of the above in place, many more people will take up urban agriculture. However, the characteristics of future farmers will basically be the same as present farmers: both the rich and poor will be involved in the practice. Moreover, low class people will continue with open-space cultivation, while middle/upper class people will mostly be involved in enclosed cultivation.

It follows also that new areas of cultivation will be added to the current ones. However, since many more people will get involved in urban agriculture, the addition of new cultivatable areas will not necessarily result in bigger areas of cultivation for the individual cultivator.

Generally, the government officials involved in this study feel the government-regulated option described above may not be probable because the Ghanaian government is increasingly becoming less interventionist. In contrast, it is encouraging private ventures with limited or no government intervention. I see it as a feasible option, especially since government involvement will be minimal, to create a favourable environment in which urban farmers operate.

I believe that the future trend of urban agriculture will be a combination of the above described options, and private commercialization of the practice as discussed below.

10.4 Private Commercialized Urban Agriculture

As already mentioned, field studies indicate that presently urban agriculture in Ghanaian towns/cities is mostly done on small scale. Thus, the relations of production is that the small-scale cultivator controls the means of production, meaning he/she consumes the surpluses that accrue from his/her labour. However, indications are that these relations will change in the future. Urban Ghana is still developing, so increasingly many plots of land are demanded for residential and other development projects. Generally, as the demand for urban land for other purposes increases, plots of land available for urban agriculture decrease, and eventually the means of production would be controlled by large-scale capitalist farm owners instead of small-scale farmers. This implies that the most economically powerful urban residents would eventually control urban agriculture, and as mentioned above, large-scale capitalist farm owners would consequently appropriate the surpluses generated by previous small-scale farmers now turned proletarians.

The demand for urban land affects open-space cultivation and enclosed cultivation differently. In line with the trend of development in Accra, and in line with modern building designs, the number of enclosed houses or compounds are increasing. This suggests that over the years more enclosed houses will be built and as a result the total land areas used for enclosed agriculture will increase. In contrast, by virtue of the fact that open-space cultivation is practiced on land earmarked for other purposes, open-space cultivators do not have a firm right of tenure. Consequently, as urban Ghana develops, land becomes less available for open-space cultivation. Therefore, the general trend is that while enclosed cultivation continues to increase, open-space cultivation will continue to grow to a certain level and then decline. It also means the number of low status cultivators will increase to a level and then decrease. This trend will finally level up when the only spaces left for open-space farming are those not suitable for other purposes. Contrarily, when the number of low status cultivators has ceased to increase, the number of middle/upper class cultivators will continue to increase.

Following this trend, there are some important questions that need to be addressed.

1. Will the sizes of farm plots increase or decrease? Principally, this depends on the rate at which land is developed for residential and other purposes. Based on current developmental trend in urban Ghana, my projection, as mentioned above, is that many plots of land will be converted into estates, factories, and the like. It follows that after a point, as the number of open-space cultivators increases, the total area of urban land available to them will diminish. As a result, many open-space cultivators will share smaller cultivable open space. This will have major implications for individual open-space cultivators for example, as their plot sizes will decrease, subsequently, their income from cultivation will decrease.

As more and more urban land is developed, the total area of cultivable land available to enclosed cultivators increases. This will happen because increasingly many middle/upper class people are building or buying enclosed houses. Although the area available to enclosed cultivators will increase, actual cultivatable space available to the individual will not increase. Here, I assume that the size of plots allocated for building houses remains the same.

2. What types of crops will be mostly cultivated? As noted in chapter 7, most vegetables sold in Accra are produced by open-space cultivators.

When the area of land available to open-space cultivators decreases, they will be unable to produce a sufficient amount for the market.[1] A logical assumption from this observation is that eventually the cultivation of vegetables will decrease. However, this does not seem to be very dramatic. As noted in chapter 7, middle/upper class people are very interested in growing vegetables. Hence, with its decrease in the market supply by open-space cultivators, they (enclosed cultivators) will devote most of their lands to the cultivation of vegetables. Consequently, in the future, the crops cultivated will not be very different from those cultivated today.

3. Will the lower class move to the periphery and rural areas to farm? According to the induced institutional innovation hypothesis,

> Changes in institutional arrangements governing the use of production factors [are] induced when disequilibria between the marginal returns and the marginal costs of factor inputs [occur] as a result of changes in factor endowments and technical change (Hayami and Ruttan, 1985:102).

Thus, loss of land and decreasing land area available to open-space cultivators should compel them to adjust their present ways of cultivation. For example, most of them would have to cultivate at the periphery of urban Ghana. Since economics is a large factor in the choice of jobs for people, lower class cultivators will move to the periphery of urban areas and/or to rural areas only if they will make profits by such a move. However, indications are that this does not seem likely.

Moving to the periphery means relocation. It costs a lot of money principally in the form of accommodation money that many in the lower class do not have. Consequently, lower class urban cultivators will rather remain at core areas and commute to the periphery to cultivate; however commuting also costs money. My point is that as open-space cultivators lose land to developers at the core of towns/cities, they would have to commute to the periphery to cultivate. This will bring additional costs to them. Whether they will be able and/or willing to commute to the periphery to cultivate depends on the profit they will eventually make.

Commuting to farm at the periphery, and therefore profits, depends on the market prices of vegetables and other crops grown by open-space

[1] It is assumed that they will not be able to increase their yield significantly with improved technology.

cultivators. In Ghana, the prices of food are influenced by transportation costs, not vice versa. And transportation costs are influenced by the price of petroleum. Considering the fact that vehicular petrol prices continue to rise, my projection is that transportation costs will continue rising, thereby forcing the prices of vegetables and other crops up. A point will be reached where the prices of vegetables, especially, will be very high. Beyond that point prices will stabilize or go down relatively while transportation costs will continue rising, thus making it less profitable for commuting open-space cultivators, especially since they cultivate on small-scale. The majority will therefore abandon commuting to the periphery to cultivate, and hence abandon their farms.[2]

Ghanaian urban residents rarely move to resettle in rural areas. Even an enticing government package failed to attract redundant urban residents to relocate in rural areas (see chapter 4). Consequently, open-space cultivators who lose their farmlands, people who will not have any enticing packages from the government, will not move to rural Ghana. They will remain in the towns/cities and shift their attention to other informal sector activities like retailing, and black marketeering, or sell their labour.[3]

4. Considering the fact that open-space cultivators grow vegetables, how will a decreasing availability of land affect the supply of vegetables in Ghanaian cities? As mentioned above, enclosed cultivators will be the ones to increase cultivation of vegetables but it should be remembered that they do not cultivate for sale. In addition, not all people who live in enclosed houses put their land to agronomic use. Some cultivate flowers and others use the land as playgrounds for their children. Consequently, for a period of time there will be a decrease in the quantity of vegetables on the market. Conversely, due to increased population and increased awareness of the importance of vegetables the demand of vegetables will actually go up.

The decrease in the supply of vegetables to the market and the increase in demand will force up the prices of vegetables. As economic beings, some middle/upper class people will take advantage of this situation. Subsequently, the middle/upper class will engage themselves in commercial production of vegetables at the periphery or outside the immediate

---

[2] Commercial farming will compel most small-scale farmers to abandon urban farming.

[3] The last option (selling their labour) is a feasible option only because of emerging commercial agriculture. Thus, they will sell their labour as farm workers.

boundaries of urban areas.

Most of the open-space cultivators who lose their farming land to the original owners will have to find other means of survival. For these people, one of the most important means to make a living is to sell their labour to the emerging urban farm capitalist (or urban commercial farmers). Therefore, the increased development (in the form of built-up areas) of Ghanaian urban areas further enhances the development of proletarianization.

5. What gender will dominate urban cultivation? My proposition is that between the period of open-space cultivators losing the land on which they farm, and the emergence of commercialization of urban agriculture in Ghana, most cultivators will be females. This is because most of those losing their land will be males. Conversely, as noted in chapter 7, most of the people involved in enclosed cultivation, those who will not lose their lands will be females. As also mentioned in chapter 7, in Ghana and other sub-Saharan African countries food production for home consumption is mostly done by women. Since enclosed cultivation is for home consumption, most of the cultivators at this period will be females.

However, most of the people who will eventually run the commercial urban farms will be men because in sub-Saharan African farming, men work to earn cash while women engage in subsistence farming (see chapter 7). Similarly, most of the people who will be hired as permanent workers on commercial farms will be men. In the first place, as mentioned, most of those who lose their farmlands are men, and consequently most of those who will be willing to sell their labour to the capitalist farmers will be men.

Secondly, as already mentioned in chapter 7, Ghanaian women prefer distribution or retailing to urban agriculture. Therefore, the majority of women will not be open to the idea of selling their labour as farm workers they would rather become retailers.

Thirdly, urban capitalist farmers will 'favour male employees because of their past experience with them and because of the supposition that the jobs are men's' (Palmer, 1991:44). Also in comparison to men, women may be seen as less reliable due to absenteeism necessitated by maternity leave. They may also be considered low producers during pregnancy.

Generally, in Ghana men will still form the bulk of urban farmers, both as farm labourers and farm owners.

It follows from the above arguments that urban cultivation, which started as a small-scale gardening, will eventually become large-scale commercialized farming. This does not entail absence of small-scale urban cultivation. Rather, the emphasis will be on commercial cultivation. Sociologically, this can be seen as a reflection of the competition between capitalists and *petit bourgeoisie* with an obvious outcome of the former absorbing the enterprises/initiatives of the latter, compelling the latter to sell their labour to the former.

10.5 Policy Recommendations

In the previous section, I noted that there will be a change of focus from open-space to enclosed cultivation (both small-scale), then to large-scale urban farming. Policies or regulations should therefore change to reflect the dominant mode of production at different periods. The regulations should be 'a trade-off... between the regulation of health and safety and the economic value of informal sector activities' (Stren et al., 1992:86), in this case urban agriculture.

At the present, when urban cultivation is on small-scale and the emphasis is on open-space cultivation, the government should primarily address the problems of open-space cultivators. First and foremost, regulations which are not enforceable should not be enacted because it is irrational to enact laws that cannot be enforced, or which the government is not interested in enforcing. For example, in Ghana it is forbidden by a by-law to sell foodstuffs on the farm. However, many consumers prefer to buy on the farms and consequently urban cultivators sell on their farms. It seems the government is not really willing to enforce this law against unharmful behaviour. However, with the existence of this by-law a corrupt official can extort money from cultivators who sell on their farms. By discarding such by-laws, the government limits the grounds on which corrupt officials harass cultivators. Now, some concrete recommendations.

1. The government should provide favourable conditions that will increase yields. However, in order to avoid government paternalism, urban farmers should not be given any preferential treatment in comparison to rural farmers. My suggestion is that amenities like credits from the Agricultural Development Bank and the services of Agricultural Extension Officers offered to rural farmers should be extended to urban farmers. An important inference from the preceding statement is that there should be stronger linkages between the formal and informal sectors.

2. Following from the above, at every potential farming area there

should be a signboard advising prospective farmers about the government organ they should seek permission from before cultivating the land. For example, if a piece of land belongs to the Ministry of Defence, a sign-post should indicate which office in the Defence ministry prospective cultivators should contact for lease of a piece of the land. Some cultivators in Accra fear that such a step will put them at the mercy of corrupt officials, however, such a measure if properly implemented will accord cultivators secure land tenure. When government organs register farmers who cultivate their land, the former will know who to contact and give information prior to retrieving the land for development.

3. Concerning private land, through incentives like tax relief, the government should encourage private landlords to rent their land to urban farmers.

4. There should be a by-law that stipulates that before a piece of land is taken away from a cultivator, he/she should be given at least three months notice, depending on the crops under cultivation. In this regard, in order to prevent conflict between cultivators and landlords, both parties should agree on the types of crops to be cultivated on the land.

5. Another way to boost urban agriculture as it is practiced today is for government organs like the Accra Metropolitan Assembly to make compost from refuse, bag it and truck it to fields to sell to cultivators. At the moment, the Accra Metropolitan Assembly makes compost but it is not aggressive in marketing it. Urban cultivators are expected to go to the composting site to purchase but most of them (the cultivators) do not have any means of transportation, and they are not inclined to spend money in carting unbagged compost to their fields because, as one of them put it, it is too bulky. Similarly, the cultivators are not in dire need of compost because they have easy access to the less bulky bags of fertilizer, which they use as substitute.

6. Furthermore, since apartment buildings are increasingly becoming common in Ghana, the government should insist on special building codes that compel estate developers to set aside plots of land behind and in between apartment blocks for cultivation. It follows therefore that contemporary sub-Saharan African towns and cities should learn from ancient African cities which 'set aside maxispace within their walled confines for food production' (Hull, 1976:395).

7. Apart from steps to increase the output of urban farmers, the government should put protection mechanisms in place to protect the health and safety of both urban farmers and consumers. On the health and safety

aspect of urban agriculture, Health Officials should go round to check and advise urban cultivators on the type of water they use.

8. If possible, gutter water should be cleaned and released to cultivators instead of the untreated gutter water they currently use.

9. Government experts should control the type of chemicals urban farmers use. As mentioned in chapter 5, some cultivators still use gammalin 20 (DDT), which is prohibited in all developed and several developing countries because of its long-lasting danger to plants, animals and humans.

10. In addition to the above steps to protect the health of urban farmers and consumers, Veterinary Officers should periodically inspect animals reared in urban areas to control animal diseases, and to control animal diseases that may be transmitted to humans. Such measures, like periodic inspection of animals, are not provided in rural areas. Consequently, one may want to know why I expect officials to provide such services to urban farmers after I have previously suggested that urban farmers should not be given preferential treatment. I recommend periodic inspection of animals and other such measures because in urban agriculture there is close proximity between animals and humans.

The above intervention methods recommended for the government will be necessary for a specific period when the emphasis on urban agriculture is still on small-scale farming, and when cultivation and animal rearing is done in close proximity to humans. However, in the future, emphasis will shift to commercial farming where farms will be further away from human concentrations. One may ask, 'if in the future less emphasis would be placed on small-scale farming, then why should the government implement the above suggested measures to improve upon this type of farming?' The suggested measures would still be necessary because small-scale farming will not vanish, thus, it would exist alongside commercial farming. Consequently, to improve upon it and to ensure that the farmers adhere to healthy farming practices it would be necessary for the government to regulate small-scale farming.

Since the emphasis would be on commercial farming, different intrusion methods as mentioned below will be necessary.

To boost commercial urban agriculture, the government should provide a healthy atmosphere for its practice. Thus, I advocate for a more limited intrusion in commercial urban farming by the government because Ghana has not fared well when it comes to her involvement in farms.

1. Commercial farmers should be taxed based on their sales, consequently, one important intrusion method is taxation, which will bring some revenue into the country. On the other hand, it will give farmers the power to demand services like technical advice and credit from the government because people who pay taxes have the right to demand some services from their government.

2. The government should construct roads to all peri-urban areas where commercial farming takes place, and where farming is likely to take place. This suggestion is made with the hope that availability of motorable roads will encourage commercial farmers to produce more, since they will be able to cart their farm products to the market. In addition, motorable roads will make it possible for them to cart agricultural equipment to their farms.

3. The success of commercial urban agriculture is partly dependent upon the availability of farm equipment at reasonable prices. Consequently, the government should encourage the manufacture of most of the agricultural equipments needed in the country, and remove levies imposed on imported agricultural equipments.

10.6 Conclusion

In this work, I mentioned that urban agriculture is very necessary for the survival of many urban residents because they do not have any formal employment, and some others are lowly waged, making it vital for them to seek extra income from farming. It is pertinent to acknowledge that despite their low income or unemployment, not all urban residents are involved in urban agriculture. Some unwaged urban residents get involved in other informal sector activities instead. In addition, some of the urban residents who are formally employed are involved in urban farming, while others are not.

I used three models to explain why some people involve themselves in urban agriculture. The models should be seen as complementary, since each of them explains a part of the whole and none of them singularly explains the whole i.e. none of the three theories explains all the reasons why some people cultivate in urban areas and others do not. For example, the labour-surplus model succeeds in explaining why many formally unemployed urban residents involve themselves in farming but it does not succeed in explaining why some formally employed urban residents are also involved in urban agriculture. This explanation, i.e. why some formally employed urban residents are involved in urban agriculture, is rendered by the dependency model of Third World economies. These two models, however, ignore a very important variable in explaining the characteristics of urban farmers. The cultural lag model brings in the rural background of urban farmers as an important factor in explaining why some urban residents get involved in farming and others do not. This model notes that many urban residents farm in urban areas because of their rural background or their previous experience in farming.

A fundamental issue is, what happens as more and more people are born and raised in Accra or urban Ghana in general? Thus, in the future, how important will the cultural background of urban residents be, as far as urban agriculture is concerned? My prediction is that, those born and raised in urban Ghana will not be interested in urban agriculture, especially as a small-scale practice. However, due to its great economic value commercial agriculture, on the other hand, will be attractive to both urban- and rural-raised Ghanaians. Furthermore, the demographic trend in urban Ghana is following that experienced in advanced countries a lower birthrate. In comparison, the birthrate in rural Ghana is still high, meaning rural Ghana will still be an important supplier of people to urban Ghana.[4] This implies that in the future there will still be many urban residents with rural backgrounds. Consequently, the importance of urban agriculture will not lessen by the fact that many Ghanaians will be born and raised in urban Ghana.

In Ghana, like some other sub-Saharan African countries, urban agriculture has had a long journey from total proscription to official accommodation. It is contended in this work that government officials inherited a negative attitude toward urban agriculture from colonial officials. Secondly, government officials did not condone urban agriculture because they did not know much about it. Thirdly, they undervalued the importance of urban agriculture as a practice that curtails urban unemployment and provides fresh vegetables for urban residents. When the magnitude of the importance of urban agriculture became clearer, government officials, at least, condoned it. In other words, urban farmers played a game of hide and seek with officials until the latter gave up.

It is important to emphasize that in Accra many people of high socio-economic status, including government officials, are now involved in the

[4] This is on the condition that the factors pulling rural residents to urban Ghana will still be in place.

practice. This means the characteristics of urban farmers have changed over the years, from almost only lower class people to all classes of people, thus raising its prestige. Increased democracy in sub-Saharan Africa also contributed to the change in official attitude toward urban agriculture. In this region, political activism is higher in the urban areas and in order to win the votes of urban residents and discourage them from embarking on industrial actions, governments have desisted from antagonising them by allowing agriculture in the towns/cities.

As mentioned in the preceding paragraph, urban farmers come from all types of classes. They are male and female, from various age groups, and are not recent migrants, meaning they are people who have adjusted to urban life.

On the basis of the location of farms or gardens, I identified two main modes or types of cultivation in Accra. These are open-space cultivation and enclosed cultivation. Open-space cultivators cultivate mostly for sale, while enclosed cultivators do so for home consumption. Men dominate the former, while women dominate the latter, meaning men dominate the economic aspect of urban agriculture. Comparatively, open-space cultivators are of lower socio-economic status.

Presently, open-space cultivation is the most common type of urban cultivation. However, in the future, due to land scarcity, its importance will dwindle, while that of enclosed cultivation will blossom, and finally the main emphasis will be on commercial agriculture. Therefore, I foresee that the emphasis will shift from open-space small-scale cultivation to commercial or large-scale cultivation, meaning the most important players in the practice will change from small-scale farmers to capitalists.

Bibliography

Acheampong, I. K. (1972). *Redeeming the Economy.* Accra: Ghana Government Press.

Adejumobi, A. (1992). *Coping Strategies for Survival among Unemployed Nigerian Graduates.* Monograph, No. 12. Lagos: NISER.

Agbodeka, Francis (1992). *An Economic History of Ghana.* Accra: Ghana University Press.

Agricultural Development Bank (1975). *Agricultural Credit Programmes in Ghana.* Accra: ADB.

Aidoo, Thomas Kwasi (1983). "Ghana: Social Class, the December Coup, and the Prospects for Socialism," in Bernard Magubane and Nzongola-Ntalaja (eds.) *Proletarianization and Class Struggle in Africa.* San Francisco: Synthesis Publications.

Alderman, Harold and Gerald Shively (1996). *Economic Reform and Food Prices: Evidence from Markets in Ghana.* World Development 24(3): 521-534.

Andrae, Gunilla (1992). "Urban Workers as Farmers: Agro-links of Nigerian Textile Workers in the Crisis of the 1980s" in J. Baker & Pal Ove Pedersen (eds.) *The Rural-Urban Interface in Africa.* Uppsala: Nordiska Afrikaintitutet.

Anyane, S. La (1963). Ghana Agriculture: *Its Economic Development from Early Times to the Middle of the Twentieth Century.* London: Oxford University Press.

Arrighi, G. and Saul, J.S. (1973). *Essays on the Political Economy of Africa.* New York: Monthly Review.

Aryeetey, Ernest (1992). *The Relationship Between the Formal and Informal Sectors of the Financial Market in Ghana.* Oxford: Centre for the Study of African Economies.

Bartone, Carl (1994). "Chile: Managing Environmental Problems: Economic Analysis of Selected Issues," Report 13061 CH. Washington, D.C.: Environment and Urban Development Division, World Bank.

Bender, David A. (1997). *Introduction to Nutrition and Metabolism*. London: Taylor and Francis.

Benneh G, Songsore J, Nabila J.S, Amuzu A.T, Tutu K.A, Yangyuoru Y, and McGranahan, G (1993). *Environmental Problems and the Urban Household in the Greater Accra Metroplitan Area (GAMA)- Ghana.* Stockholm: Stockholm Environment Institute.

Bigsten, Arne and Kayizzi-Mugerwa, S. (1992). "Adaptation and Distress in the Urban Economy: A Study of Kampala Households," World Development Vol. 20(10).

Binns, Tony (1994). *Tropical Africa.* New York: Routledge.

Blau, Peter (1986). "Microprocess and Macrostructure", in Karen Cook (ed.) Social Exchange Theory. Newbury Park: Sage Publications.

Blau, Peter and Schwartz, Joseph (1984). *Crosscutting Social Circles.* Orlando, FL.: Academic Press.

Bowditch, Edward (1819). *Mission from Cape Coast Castle to Ashantee with a Statistical Account of that Kingdom*. London: HMSO.

Braun, J, McComb, J., Fred-Mensah, B., and Pandya-Lorch, R. (1993). *Urban Food Insecurity and Malnutrition in Developing Countries.* Washington, D.C.: International Food Policy Research Institute.

Broom, L., Selznick, P., and Darroch, D. B. (1981). *Sociology. A Text with Adapted Readings.* New York: Harper and Row Publishers.

Brown, C.K. (1993). "Socio-cultural Aspects or Rural Credit in Ghana" Paper presented at the Workshop on Rural Finance at Ghana Institute of Management and Public Administration, Accra on 24-26 Feb., 1993.

Bryceson, Deborah F. (1987). "A Century of Food Supply in Dar es Salaam", in Jane I. Guyer (ed.) Feeding African Cities: Studies in *Regional Social History.* Manchester: Manchester University Press.

Cadwell, J.C. (1968). *Population Growth and Family Change in Africa: The New Urban Elite in Ghana.* Canberra: Australian National University Press.

Castells, Manuel and Alejandro Portes (1989). "World Underneath: The Origins, Dynamics, and Effects of the Informal Economy," in Alejandro Portes, Manuels Castells, and Lauren A.Benton (eds). *The Informal Economy*. Baltimore: The Johns Hopkins University Press.

Central Bureau of Statistics (1970). *Ghana: Economic Survey 1966-69.* Accra: Ministry of Information.

Chimhowu, Admos and Gumbo, D. (1993). "Urban Agriculture: Southern and Eastern Africa", in Luc Mougeot and D. Masse (eds). *Urban Environment. Vol. 1.* Ottawa: IDRC.

Commission on the Civil Service (1951). *Report of the Commission on the Civil Service of the Gold Coast 1950-51.* Accra: Government Publishers.

Curtis, John W. (1995). *Opportunity and Obligation in Nairobi. Social Networks and Differentiation in the Political Economy of Kenya.* Bayreuth: LIT Verlag.

Diallo, S. (1993). "A Plot of One's Own in West African Cities", in *Farming in the City: The Rise of Urban Agriculture.* Reports Vol. 21, No.3. Ottawa: IDRC.

Dettwyler, Steven P. (1985). *Senoufo Migrants in Bamako: Changing Agricultural Production Strategies and Household Organization in an Urban Environment.* Ph.D (Thesis) Indiana University.

Drakakis-Smith, D. (1992). *Food Production and Under-nutrition in Third World Cities*. Hunger Notes, 18(1):5).

Duncan, Beatrice Akua (1997). *Women in Agriculture in Ghana.* Accra: Gold-Type Ltd.

Egziabher Axumite G. (1994). "Urban Farming, Cooperatives, and the Urban Poor in Addis Ababa" in Luc Mougeot et.al (eds.) *Cities Feeding People: An Examination of Urban Agriculture in East Africa*. Ottawa: IDRC.

Ekeh, Peter (1974). Social Exchange Theory. London: Heinemann Educational Books.

Ewusi, Kodwo (1986). Industrialization, Employment Generation and income Distribution in Ghana, 1950-1986. Accra: Adwensa Publications.

FAO (1987). 1948-1985 World Crop and Livestock Statistics. Rome: FAO.

—— (1968). Production Yearbook 1967. Rome: FAO.

Fapohunda, O.J., Adebgola O. & Pius Sada (1986). Population, Employment and Living Conditions in Lagos. Ibadan: University Press Limited.

Fashoyin, Tayo (1994). "Preface," to Tayo Fashoyin (ed.) *Economic Reform Policies and the Labour Market in Nigeria.* Lagos: NIRA.

Finance, Ministry of (1958). *Ghana: Economic Survey 1957.* Accra Government Printing Department.

Freeman, Donald B. (1991). *A City of Farmers: Informal Urban Agriculture in the Open Spaces of Nairobi, Kenya.* Montreal: McGill-Queen's University Press.

Gibbon, Peter (1993). "Introduction: Economic Reform and Social Change in Africa," in Peter Gibbon (ed.) *Social Change and Economic Reform in Africa.* Uppsala: Nordiska Afrikainstitutet.

Gibson, K. (1980). "The Internationalisation of Capital and Uneven Development within Capitalist Countries" in R. Peet (ed) *An Introduction to Marxist Theories of Underdevelopment.* Canberra: Department of Human Geography, Australian National University.

Godfrey, E.M. (1973). "Economic Variables and Rural-Urban Migration: Some Thoughts on the Todaro Hypothesis," *Journal of Development Studies, Vol.10, No.1.*

Ghana Statistical Service (1994). *Ghana Demographic and Health Survey 1993.* Accra: Ghana Statistical Service.

——— (1995). *The Pattern of Poverty in Ghana 1988-1992.* Accra: Ghana Statistical Service.

Globe and Mail (1994). "Urban Cowboy Milks Profits on Small Spread," Feb. 2.

Goodland, J.A., Watson, C., & Ledec, G. (1984). *Environmental Management in Tropical Agriculture.* Boulder, CO: Westview Press.

Gottdiener, Mark (1994). *The New Urban Sociology.* New York: McGraw-Hill, Inc.

Granovetter, Mark S.(1973). "The Strength of Weak Ties," in *American Journal of Sociology, Vol.78(6).*

Gregory, D. and Urry, J. (1985). *Social Relations and Spatial Structures.* London: Macmillan.

Grigg, David (1995). *An Introduction to Agricultural Geography.* New York: Routledge.

Gutkind, Peter (1963). *The Royal Capital of Buganda: A Study of Internal Conflict and External Ambiguity.* The Hague: ISS.

Guyer, J.I. (1987). *Feeding African Cities: Studies in Regional Social History.* Manchester: Manchester University Press.

Gyimah-Boadi, E. (1989). "Policies and Politics of Export Agriculture", in E. Hansen and K.A. Ninsin (eds) *The State, Development and the Politics of Ghana.* London: Codesria.

Hammer, Muriel (1983). "Core and Extended Social Networks in Relation to Health and Illness." *Social Science and Medicine, No. 17.*

Hansen, Emmanuel (1989). "The State and Food Agriculture", in E. Hansen and K. A. Ninsin (eds.) *The State, Development and the Politics of Ghana.* London: Codesria.

Hardin, G.J. (1972). *Exploring New Ethics for Survival.* New York: Viking.

Hayami, Yuhiro and Ruttan Vernon (1985). *Agricultural Development: An International Perspective.* Baltimore:The John Hopkins University Press.

Hellevik, Ottar (1984). *Introduction to Causal Analysis: Exploring Survey Data by Crosstabulation.* London: George Allen & Unwin.

Holm, Morgens (1992). "Survival Strategies of Migrants to Makambako- an Intermediate Town in Tanzania," in Baker, J. and Pedersen, Pal Ove (eds.) *The Rural-Urban Interface in Africa.* Uppsala: Nordiska Afrikainstitutet.

House, William (1978). "The Urban Informal Sector:Its Potential for Generating Income and Employment Opportunities in Kenya." *Occasional Paper No.25.* Nairobi: Institute for Development Studies, University of Nairobi.

Hull, Richard W. (1976). "Urban Design and Architecture in Precolonial Africa," in the *Journal of Urban History,* Vol.2, No.4, Beverly Hills, California: Sage Publications.

Hussain, M.A. and P. Lunven (1989). "Urbanization and Hunger in Cities," in *Food and Nutrition Bulletin, 9 (4).*

Ikpeze, Nnaemeka (1994). "The Effects of Structural Adjustment Programme on Employment in Nigeria's Large and Medium size Establishments," in Tayo Fashoyin (ed.) *Economic Reform Policies and the Labour Market in Nigeria.* Lagos: NIRA.

ILO (1991). "The Dilemma of the Informal Sector," *Report of the Director-General.* Geneva: ILO.

—— (1991b). *African Employment Report 1990.* Addis Ababa: JASPA.

IMF (1996). *World Economic Outlook May 1996.* Washington, D.C.: IMF.

ISSER (1995). *The State of the Ghanaian Economy, 1994*. Legon: ISSER.

Jacobs, J. (1970). *The Economy of Cities*. London: Jonathan Cape.

Kraus, J. (1986). "The Political Economy of Agrarian Regression in Ghana," in Cummins, Stephen et. al. (eds.) *Africa's Agrarian Crisis: The Roots of Farming*. Boulder, CO: Lynne Reiner Publishers.

Knight, J.B. (1972). "Rural-urban Income Comparisons and Migration in Ghana", in *Bulletin, Oxford University Institute of Economics and Statistics, 34 (2):199-228.*

La Anyane, S. (1963). *Ghana: Agriculture. Its Economic Development from Early Times to the Middle of the Twentieth Century.* London: Oxford University Press.

Lamba, D. (1993). "The Neglected Industries of Kenyan Cities", in *Farming in the City: The Rise of Urban Agriculture. Reports Vol.21, No.3.* Ottawa: IDRC.

Lawson, Rowena (1963). "A Human Needs Diet. The Western Pattern in Ghana", in *The Economic Bulletin of Ghana, 7, (1):35-36.*

Lawson, Rowena (1967). "The Distributive System in Ghana. A Review Article", in *Journal of Development Studies, 3, (3):195-205.*

Lee-Smith, D; Manundu, M; Lamba, D; Gathuru, K (1987). *Urban Food Production and the Cooking Fuel Situation in Urban Kenya- Nairobi Report",* Nairobi: Mazingira Institute.

Lee-Smith, D. & Lamba, D. (1991). "The Potential of Urban Farming in Africa", in Ecodecision. December, Montreal.

Lee-Smith, D. & Memon P. A. (1994). "Urban Agriculture in Kenya," in L.A. Mougeot (ed.) *Cities Feeding People.* Ottawa: IDRC.

Leys, Colin (1975). *Underdevelopment in Kenya: The Political Economy of Neo-Colonialism.* London: Heinemann.

Lowder, Stella (1986). *Inside Third World Cities*. London: Croom Helm.

Mabogunje, Akin (1986). "Backwash Urbanization: The Peasantization of Cities in Sub-Saharan Africa," in Michael Conzen (ed.) *World Patterns of Modern Urban Change.* Chicago: The University of Chicago Press.

March, J.G. and Shapira, Z. (1992). "Behavioral Decision Theory and Organizational Decision Theory," in Mary Zey (ed.) *Decision Making: Alternative to Rational Choice Models*. London: Sage Publications.

Maxwell, D & Zziwa, Samuel (1992). *Urban Farming in Africa.* Nairobi: ACTS.

Maxwell, D. (1993). "Farming Logic in Kampala," in *Farming in the City: The Rise of Urban* Agriculture. Reports Vol.21, No.3. Ottawa: IDRC.

——— (1994). "The Household Logic of Urban Farming in Kampala", in Luc Mougeot et al. (eds.) *Cities Feeding People: An Examination of Urban Agriculture in East Africa.* Ottawa: IDRC, 1994.

——— (1997). "The Political Economy of Urban Food Security in Sub-Saharan Africa," paper presented to the International Conference on Sustainable Urban Food Systems on May 22-25, Ryerson Polytechnic University, Toronto, Canada.

Mazingira Institute (1987). *Urban Food Production and the Cooking Fuel Situation in Urban Kenya.* Nairobi: Mazingira Institute.

Mbiba, Beacon (1995). *Urban Agriculture in Zimbabwe.* Aldershot: Avebury.

Michelson, William (1976). *Man and his Urban Environment: A Sociological Approach.* Toronto: Addison-Wesley Publishing Company.

Mittar, Vishwa (1988). *Growth of Urban Informal Sector in a Developing Economy*. New Delhi: Deep and Deep Publications.

Mitchell, J.C. (1987). *Cities, Society and Social Perception: A Central African Perspective*. Oxford: Clarendon.

Mitchell, J.C. (1969). "The Concept and use of Social Networks," in J.C. Mitchell (ed.) *Urban Situations.* Manchester: Manchester University Press.

Mlozi, M.R.S., Lupanga, I.J. & Mvena, Z.S.K. (1992). "Urban Agriculture as a Survival Strategy in Tanzania" in J. Baker and P.O. Pedersen (eds.) *The Rural-Urban Interface in Africa.* Uppsala: Nordiska Afrikainstitutet.

Morrison, Thomas (1984). "Cereal Imports by Developing Countries. Trends and Determinants," in *Food Policy, 9(1):13-26.*

Moser, C.O.N. (1984). "The Informal Sector Reworked: Viability and Vulnerability in Urban Development," in *Regional Development Dialogue, 5(2): 135-179.*

——— (1978). "Informal Sector or Petty Commodity Production: Dualism or Dependence in Urban Development?," in *World Development 6, No.9/10: 1041-1064.*

Mosha, A.C. (1991). "Urban Agricultural Activities in Tanzania," in *Review of Rural and Urban Planning in Southern and Eastern Africa, Vol. 1.*

Mougeot, Luc (1993). "Urban Food Self-Reliance: Significance and Prospects," in *Farming in the City: The Rise of Urban Agriculture.* Reports Vol.21, No.3. Ottawa: IDRC.

Mtwere, M (1987). *The Major Bottlenecks Faced by Smallholder Pig Producers in the Areas Surrounding Dar es Salaam.* Morogoro: Sokoine University of Agriculture.

Mvena, Z.S.K., Lupanga, I.J. & Mlozi, M.R.S. (1991). *Urban Agriculture in Tanzania: a Study of Six Towns.* IDRC Report 86-0090. Ottawa: IDRC.

Ngwa, Nebasina E. (1987). "Time and Land Space Utilization within an Urban Confine: The Case of Buea Town Gardeners in the Republic of Cameroon," *in GeoJournal, Vol.15, No.1.*

Obosu-Mensah, Kwaku (1996). *Ghana's Volta Resettlement Scheme: The Long-Term Consequences of Post-Colonial State Planning*. Bethesda: International Scholars Publishers.

Ohadike, Patrick (1988). *Development in Africa.* Legon: Woeli Publishing Services.

Oke, E.A. (1986). "Kinship Interaction in Nigeria in Relation to Societal Modernization: A Pragmatic Approach," in *Journal of Comparative Family Studies, 17(2), 185-196.*

Oloya, C. (1988). "Some Aspects of Rural Way of Life in an Urban Setting: A Case Study of Kampala City," B.Sc. Thesis, Dept. of Geography. Kampala: Makerere Uni. Cited in Maxwell and Zziwa's (1992) *Urban Farming in Africa: The Case of Kampala, Uganda.* Nairobi: ACTS.

Oni, Bankole (1994). "SAP and the Informal Sector in Nigeria: a Case of Increasing Employment," in Tayo Fashoyin (ed.) *Economic Reform Policies and the Labour Market in Nigeria*. Lagos: NIRA.

——(1987). "The Structural Adjustment Programme and Unemployment: an Evaluation," in Adedotun Phillips and Eddy Ndekwu (eds.) *Structural Adjustment Programme in a Developing Economy.* Lagos: NISER.

Oppong, Christine (1992). "Traditional Family Systems in Rural Settings in Africa," in Elza Berquo and Peter Xenos (eds.) *Family Systems and Cultural Change.* Oxford: Oxford University Press.

Orde-Browne, G. St. J. (1926). *Labour in the Tanganyika Territory.* London: HMSO

Oucho, John and W.T.S. Gould (1993). "Internal Migration, Urbanization, and Population Distribution," in Karen Foote, Kenneth Hill and Linda Martin (ed.) Demographic Change in Sub-Saharan Africa. Washington, D.C.: National Academy Press.

Palmer, Ingrid (1991). *Gender and Population in the Adjustment of African Economies: Planning for Change*. Geneva: ILO.

Pinstrup-Andersen, Per (1994). "The Food Situation in sub-Saharan Africa and Priorities for Food Policy and Donor Assistance," in *Food Policy in sub-Saharan Africa: A New Agenda for Research and Donor Assistance.* London: NRI/IFPRI.

Rakodi, Carole (1988). "Urban Agriculture: Research Questions and Zambian Evidence", in *Journal of Modern African Studies 26, No.3:495-515.*

——(1997). "Global Forces, Urban Change, and Urban Management in Africa," in Carole Rakodi (ed.) *The Urban Challenge in Africa.* Tokyo: United Nations University Press.

Ratta, Annu (1993). "City Women Farm for Food and Cash," in *International Ag-Science, Vol.VI, (2).*

Rau, Bill (1993). From Feast to Famine. London: Zed Books.

Reusse, E. & R. Lawson (1969). "The Effect of Economic Development on Metropolitan Food Marketing- a Case Study of Food Retail Trade in Accra," in *East African Journal of Rural Development, 2 (1):35-55.*

Robertson, Claire C. (1984). *Sharing the Same Bowl. A Socioeconomic History of Women and Class in Accra, Ghana.* Bloomington: Indiana University Press.

Rogers, Alisdair and Steven Vertovec (1995). "Introduction" to Alisdair Rogers and Steven Vertovec (eds.) *The Urban Context: Ethnicity, Social Networks and Situational Analysis.* Washington, D.C.: Berg Publishers.

Rothchild, Donald (1991). "Overview of Structural Adjustment," in Donald Rothchild (ed.) *Ghana: the Political Economy of Recovery.* Boulder, CO: Lynne Rienner Publishers.

Sachs, I. and D. Silk (1987). "Introduction: Urban Agriculture and Self-reliance," in *Food and Nutrition Bulletin, 9 (2):2-7.*

Sandbrook, Richard (1977). "The Political Potential of African Urban Workers", in *Canadian Journal of African Studies,11(3):411-433.*

Sanyal, B. (1986). "Urban Cultivation in East Africa: People's Response to Urban Poverty," in *Food-Energy-Nexus Programme.* Tokyo: United Nations University.

Sawio, Camillus (1993). "Breaking New Ground in Dar es Salaam," in *Farming in the City: The Rise of Urban Agriculture.* Reports Vol.21, No.3. Ottawa: IDRC.

—— (1993). "Urban Agriculture in Dar es Salaam: Who are the Urban Farmers." A paper prepared for the 22nd Conference of CAAS/ACEA, Toronto, May 12-14.

—— (1994). "Who Are the Farmers of Dar es Salaam?," in Luc Mougeot et. al (eds.) *Cities Feeding People: An Examination of Urban Agriculture in East Africa.* Ottawa: IDRC.

Sen, Amartya (1990). "Food, Economics, and Entitlements," in Jean Dreze, and Amartya Sen (eds.) *The Political Economy of Hunger, Vol. 1.*Oxford: Clarendon Press.

Scott, Richard W. (1981). *Organizations: Rational, Natural and Open Systems.* Englewood Cliffs: Prentice-Hall.

Simon, David (1992). *Cities, Capital and Development: African Cities in the World Economy*. London: Belhaven Press.

Singh, Surjit (1994). *Urban Informal Sector*. New Delhi: Rawat Publications.

Sobhan, Rehman (1990). "The Politics of Hunger and Entitlement," in Jean Dreze and Amartya Sen (eds.) *The Political Economy of Hunger, Vol 1.* Oxford :Clarendon Press.

Stolka, Jiri (1987). "A Few Facts about the Hidden Economy," in S. Alessandrini and B. Dallago (eds.) *The Unofficial Economy.* Aldershot: Gower.

Stren, Richard, et al (1992). *An Urban Problematique: The Challenge of Urbanization for Development Assistance.* Toronto: University of Toronto Press.

Svedberg, Peter (1991). *Poverty and Undernutrition in Sub-Saharan Africa: Theory, Evidence, Policy.* Stockholm: Institute for International Economic Studies.

Tepperman, L. and M. Rosenberg (1995). *Micro/Micro: A Brief Introduction to Sociology*. Scarborough: Prentice Hall.

Thornton-White, L.W., L. Silberman, and P.R. Anderson (1948). *Nairobi: Master Plan for a Colonial Capital.* London: HMSO.

Tinker, Irene (1994). "Urban Agriculture is already Feeding Cities," in Mougeot, Luc et al. (eds.) *Cities Feeding People*. Ottawa: IDRC.

—— (1997). *Street Foods: Urban Food and Employment in Developing Countries.* New York: Oxford University Press.

Tinker, I. and Freidberg, S (1992). "The Invisibility of Urban Food Production," *Hunger Notes, 18(2), 3*.

Todaro, M.P. (1981). *Economic Development in the Third World.* 2nd Ed. New York: Longmans.

Tricaud, P-M. (1988). "Urban Agriculture in Ibadan and Freetown," in *Food Energy-Nexus Programme.* Tokyo: United Nations University.

U.N. (1992). *Inventory of Population Projects in Developing Countries Around the World 1991/92.* New York: U.N.

—— (1995). *World Economic and Social Survey 1995*. New York: U.N.

U.N.D.P. (1996). *Urban Agriculture: Food, Jobs and Sustainable Cities.* New York: UNDP.

UNESCO (1996). 1995 Statistical Yearbook. Paris: UNESCO.

—— (1997). 1996 Statistical Yearbook. Paris: UNESCO.

Upton, Martin (1996). *The Economics of Tropical Farming Systems.* New York: Cambridge University Press.

Vandemoortele, Jan (1991). "Labour Market Informalisation in sub-Saharan Africa," in Guy Standing and Victor Tokman (eds.) *Towards Social Adjustment. Labour Market Issues in Structural Adjustment.* Geneva: ILO (81-113).

Vennetier, P (1967a). *Les Villes d'Afrique Tropicale*. Paris: Masson.

Vennetier, P (1961). "La Vie Agricole Urbaine a Pointe-Noire," in *Cahiers d'Outre-Mer, 14(53).*

Waters-Bayer, Ann (1995). "Animal Farming in African Cities," in *African Urban Quarterly*.

Watts, T. and Bransby-Williams W. (1978). "Do Mosquitoes Breed in Maize Plants Axils?," in *Medical Journal of Zambia, Vol. 12(4).*

Webster, Andrew (1991). *Introduction to the Sociology of Development.* London: MacMillan.

Wekwete, Kadmiel (1993). "Urban Agriculture: Southern and Eastern Africa," in Luc Mougeot and Denis Masse (eds.) *Urban Environment Management. Vol. 1*. Ottawa: IDRC.

Wellman, Barry and Wortley, Scott (1989). *Different Strokes from Different Folks: Which Type of Ties Provide What Kind of Social Support?* Toronto: Centre for Urban and Community Studies.

West Africa Weekly of 24 Dec. 6 January (1991). "Re-engagement Warning." London: Benham & Co Ltd.

White, L. (1983). "A Colonial State and an African Petty Bourgeoisie,Prostitution, Property and Class Struggle in Nairobi 1936-1940," in F. Cooper (ed.) *Struggle for the City: Migrant Labour, Capital and the State in Urban Africa.* Beverly Hills, CA.: Sage Publications.

Winters, Christopher (1983). "The Classification of Traditional African Cities", in the *Journal of Urban History (Beverly Hills), 1: 3-31.*

—— (1982). "Urban Morphogenesis in Francophone Black Africa," in *Geographical Review, 72: 139-154.*

World Bank (1981). *Accelerated Development in Sub-Saharan Africa: An Agenda for Action (the Berg Report).* Washington, D.C.: World Bank.

World Commission on Environment and Development (1987). *Food 2000: Global Policies for Sustainable Agriculture.* London: Zed Books Ltd.

World Resources Institute (1992). *World Resources 1992-1993: A Guide to the Global Environment*. Oxford: Oxford University Press.

—— (1994). *World Resources 1994-1995. A Guide to the Global Environment*. Oxford: Oxford University Press.

Yeung, Yue-man (1993). "New Challenges for China's Urban Farmers," in *Farming in the City: The Rise of Urban Agriculture.* Reports Vol.21, No.3. Ottawa: IDRC.

Index